D0828937

This book of essays is a largely nonmathematical account of some of the strange behaviour, both classical and quantum, exhibited by moving particles, fluids and waves.

Drawn from the author's researches in quantum mechanics, atomic and nuclear physics, electromagnetism and optics, gravity, thermodynamics, and the physics of fluids, the essays describe different physical systems whose behaviour provokes surprise and challenges the imagination. There are strange processes for which no visualisable mechanism can be given; processes that seem to violate fundamental physical laws, but which in reality do not; processes that are superficially well understood, yet turn out to be subtly devious. The essays address questions or controversies from whose resolution emerge lessons of general significance regarding the mystery and fascination of motion. For example, does an atomic electron move? Would an 'anti-atom' fall upward? Could a particle be affected by a magnetic field that is not there? How is it possible for randomly emitted particles to arrive at a detector preferentially in pairs? Can one influence electrons in London by *not* watching them in New York? Does a Maxwell demon exist? If not, how is one to explain a simple hollow tube that blows hot air out one end and cold air out the other? Unusual in its choice and range of topics, this book reflects the unusual background and perspective of the author, a physicist and chemist whose research embraces both experiment and theory.

Anyone with a basic physics background or with an interest in the fundamental questions of physics, especially undergraduate and graduate physics students, will find this book of use.

AND YET IT MOVES

AND YET IT MOVES: STRANGE SYSTEMS AND SUBTLE QUESTIONS IN PHYSICS

MARK P. SILVERMAN

Trinity College, Connecticut

CAMBRIDGE
UNIVERSITY PRESS

Published by the Press Syndicate of the University of Cambridge
The Pitt Building, Trumpington Street, Cambridge CB2 1RP
40 West 20th Street, New York, NY 10011–4211, USA
10 Stamford Road, Oakleigh, Melbourne 3166, Australia

© Cambridge University Press 1993

First published 1993

Printed in Great Britain at the University Press, Cambridge

A catalogue record for this book is available from the British Library

Library of Congress cataloguing in publication data

Silverman, Mark P.
   And yet it moves : strange systems and subtle questions in physics
/ Mark P. Silverman.
      p. cm.
   Includes index.
   ISBN 0–521–39173–3. —ISBN 0–521–44631–7 (pbk.)
   1. Physics. 2. Quantum theory. I. Title.
QC21.2.S47   1993
530—dc20   92–38729   CIP

ISBN 0 521 39173 3 hardback
ISBN 0 521 44631 7 paperback

WV

Cooper Union Library

JUN 1 4 1994

To
Susan, Chris and Jennifer
and in memory of
H.B.S. and S.H.S.

Nature uses only the longest threads to weave her patterns, so each small piece of her fabric reveals the organization of the entire tapestry.

R. P. Feynman
The Character of Physical Law

Contents

Preface
The fire within

As a child, for as far back as I can recall, I always wanted to be a physicist – a nuclear astrophysicist or cosmologist, in fact. I am not exactly sure why, for I knew no one like that within my family or circle of acquaintances. I suspect that aspiration was owed largely to Eddington, Hoyle, Jeans and especially Gamow whose popular books I read avidly. Only rarely have the students whom I have taught over the past quarter century heard of these people or of their books. Sometimes, out of sheer perversity, or perhaps genuine curiosity, I would remark to a class, 'What? You mean you never read *One, Two, Three, . . . Infinity* or *Mr. Tompkins in Wonderland?*' But the students would only look at one another with wry smiles, as if to confirm their suspicions that physicists are strange people, and that they, unfortunately, got stuck with an especially peculiar one. Wonderland, indeed!

I never became an astrophysicist or cosmologist. Perhaps those same books which fired my imagination with the marvels of the physical world may have also led me to believe that the most fundamental mysteries of physics were largely exhausted.

I began my scientific odyssey in the field of medicine as part of a group researching malfunctions of the immune system. Finding experimentation on animals personally distasteful, however, and myself little inclined to constant preoccupation with disease, I changed to biochemistry with the heady, though misguided, notion of answering the question posed by Schrödinger's influential book, *What is Life?* It was a profound disappointment, therefore, to end up on a project to analyse the nitrite content of corn.

I took up organic chemistry instead. Predicting the outcome of complex chemical reactions by flipping electrons around pentagonal and hexagonal rings had a certain aesthetic appeal to me – at least on paper. In reality, however, 'Molecules do what they damn please!', as one professor told my wife when she was a graduate student at Harvard some years later. Having passed unscathed through more syntheses with toxic and explosive precursors than I now care to remember, I decided one day to push my luck no longer. Disaffected with a field that seemed to lack fundamental principles and in which I, quite literally, saw no future for myself, I turned to physical chemistry.

I enjoyed physical chemistry for a while, investigating molecules with electron and nuclear magnetic resonance, until I realised what it was about the subject that interested me most. It was physics. So I switched one last time and returned to the passion of my youth. I have remained a physicist ever since, and have no regrets at all for the circuitous path that finally brought me back – or almost back – to the career I decided upon as a child. If anything, the diversity of experiences has made me a better scientist.

Even during the years of 'wandering', before I rediscovered what it was I wanted to do with my life, I never actually abandoned the study of physics. I took physics courses at university, although I have little recollection of anything noteworthy about them. It was not that these courses were necessarily ill taught. At best they conveyed well enough the mathematical or mechanical skills required for solving physics problems. But something essential was missing. No instructor ever addressed the question of why physics problems were worth solving, or what made physics sufficiently interesting so that one would want to study it at all, let alone devote a lifetime to it. Not once prior to graduate school – and even then only rarely – can I recall a professor expressing personal interest in the abstractions on the classroom blackboard or the apparatus used for demonstrations. Sometimes I wonder how many potential physicists may have perished in lecture halls of universities and colleges for want of a larger vision of what physics was all about.

Fortunately for me, I did not need to rely on formal instruction for motivation. I loved to learn although I did not particularly care to be taught – at least not in the traditional manner of lecturing and testing that deprived a person of the pleasure of discovery. I already knew from childhood many reasons why physics was interesting. I needed only to know that there were still wonderful things to learn and to do, and this I gradually discovered in the same way as before, by reading widely.

My vision of physics took shape under the tutelage of Galileo, Newton, Fresnel, Maxwell, Einstein, Bohr, Born, Heisenberg, Schrödinger,

Fermi, Dirac, and a score of others whose writings I struggled through. It was perhaps not the most efficient way to learn, for there was much I did not understand until much later; at times, I understood nothing at all. But what I did absorb was priceless: a sense that in those written words and mathematical relations were ideas of fundamental importance – deep ideas that with further effort I would one day be able to comprehend. The pages spoke as if the authors, themselves, were present. In this way my passion for physics survived tiresome and seemingly pointless classroom analyses of falling projectiles, rolling cylinders and swinging pendula that many a hapless student bore somnolently to the end of his final exam – and then promptly forgot.

* * * * * * * * * * * *

What kind of physicist did I finally become? In an age when science is infinitely fragmented, its practitioners highly specialised, and experimentalists and theoreticians likely to find themselves on different floors, if not in altogether separate buildings, I hope it will appear neither coy nor audacious to give the reply I. I. Rabi gave when asked to classify himself: 'I am just a physicist'[1]. The German chemist Wilhelm Ostwald, who was much interested in the subject of scientific creativity, divided scientists into classicists, who systematically bring to perfection one or a few discoveries, and romanticists, who pursue a multitude of ideas, albeit incompletely. I rather like the colourful and sympathetic distinction drawn between these two dispositions by educator Gilbert Highet in his book, *The Immortal Profession*[2]

Will you decide (as Swift put it) to resemble a spider, spinning out endless webs from its own vitals, or a bee, visiting flower after flower and extracting a different sweetness from each of them? Will you be like those individualists one sees out west in Colorado and Wyoming, who dig their own little vertical mine shafts into the earth, and spend the rest of their days extracting ore from the same small vein? Or will you be a wandering prospector, trying first this mountain range and then that, never working out a single lode but always adventuring farther forward?

One has but to scan the employment notices in science periodicals to realise in an instant what type of scientist is sought the most today by academia, industry or government. Nevertheless, for what it is worth, I confess unabashedly to being a romanticist who has spent years happily adventuring in whatever 'mountain range' I found interesting. Unbound

[1] J. S. Rigden, *Rabi* (Basic Books, New York, 1987) 8.
[2] G. Highet, *The Immortal Profession* (Weybright and Talley, New York, 1976) 62–3.

to any one field or to any one machine, I am attracted by problems, whether of an experimental or theoretical nature, that are conceptually intriguing, even if at the time I believe I am alone in thinking so.

The essays in this book are based on some of the research with which I have instructed and entertained myself over the last couple of decades. Touching on topics drawn from quantum mechanics, atomic and nuclear physics, electromagnetism and optics, gravity, thermodynamics, and the mechanics of fluids, these essays are about different physical systems whose behaviour has stimulated my curiosity, provoked in me surprise, and challenged my imagination. There are strange processes for which no visualisable mechanism can be given; processes that seem to violate fundamental physical laws, but which in reality do not; processes that are superficially well understood, yet turn out to be subtly devious. The essays address specific questions or controversies from whose resolution emerge lessons of general significance.

For example, does an atomic electron move? How would one know? Would an 'anti-atom' fall upward? Is the vacuum really empty? Can an atom be larger than a blood cell? If it were, would it behave like a miniature planetary system? Can a particle be influenced by an electric or magnetic field that isn't there – that is, through which it does *not* pass? How is it possible for randomly emitted particles to arrive preferentially in pairs at a detector – or, conversely, to avoid one another altogether? Could watching decaying atoms emit light in London have an effect on the corresponding radiative decay in New York? Does a 'right-handed' light beam interact differently with matter than a 'left-handed' light beam? How can light get brighter by rebounding from a surface (without violating the conservation of energy)? Is a basketball changed for having being turned 360°? Perhaps not, but what about an electron? Could one tell the difference between an electron that has jumped out of a quantum state and then back again and an electron that has never jumped at all? Is there really such a thing as a 'Maxwell demon'? No? – then how is one to account for a simple hollow tube that blows hot air out one end and cold air out the other?

Broadly regarded, there is a common theme that runs through the various chapters: the mystery and fascination of motion – whether it be the movement of an electron, the flow of air, or the propagation of a light wave. It is the strange behaviour of what are often enough more-or-less familiar systems – at least to physicists – that recalls to me the famous words of Galileo ('Eppure si muove') adopted as the title of this book – words that here signify not a mutter of defiance, but rather an expression of wonder and awe.

Although largely nonmathematical, the book is not intended to be a popularisation of any aspect of contemporary physics. Nor is it designed to be a textbook or monograph. I hope, of course, that the reader may find the collection of essays instructive, but my objective is not so much to teach physics as to communicate, through discussion of personally meaningful investigations, that the study of physics can be intensely interesting and satisfying even when one is not addressing such ultimate questions as the origin and evolution of the universe. What follows, then, is essentially one scientist's personal odyssey in physics.

Admittedly, it may seem somewhat presumptuous to believe that one's own work would necessarily interest and instruct others, and for the encouragement to think thus and to see this project through to completion, I have friends and colleagues throughout the world to thank.

Indeed, one of the strongest impressions that a life in science has made upon me is the transcendence of common scientific interests over national, ethnic, racial and religious differences that somehow seem to pose such barriers to social intercourse in other walks of life. I am often reminded of Sir Humphry Davy's and Michael Faraday's peregrination through France, visiting French laboratories and factories and meeting with French scientists, at a time when France was convulsed with war. 'It is almost impossible for an inhabitant of the twentieth century to believe', wrote Faraday's biographer L. Pearce Williams[3], 'that a party of English citizens could go about their ordinary affairs in the middle of an empire locked in a struggle to the death with England without the slightest inconvenience'.

But I believe it. Under circumstances less dramatic perhaps, but nonetheless evocative of the experiences of Davy and Faraday, I have myself gone to my mailbox more than once to find – for example, from the Soviet Union and Eastern European nations during the 'cold war' or from Iran during the 'hostage crisis' – a friendly letter opening up a scientific dialogue or extending an invitation to visit and lecture. Where else, but in science, I have often thought, would it be so natural and proper for total strangers half a world apart to exchange letters telling of their deepest interests. More than one scientific adventure began, in fact, with my opening or writing such a letter.

In this regard, it is a pleasure to acknowledge the benefit of many delightful, far-ranging conversations and shared experiences with four colleagues in particular: Professor Jacques Badoz of the Ecole

[3] L. Pearce Williams, *Michael Faraday* (Simon and Schuster, New York, 1971) 36.

Supérieure de Physique et Chimie Industriclles (Paris), Professor Geoffrey Stedman of the University of Canterbury (Christchurch), and Dr Akira Tonomura and Dr Hiroshi Motoda of the Hitachi Advanced Research Laboratory (Tokyo). It is not only those kinds of motion prescribed by physical laws that elicit wonder, but also, in a metaphorical sense, the extraordinary exchange of ideas and people that characterise the scientific enterprise itself.

To someone like me, who has been for most of his professional life simultaneously a physicist and a teacher, the pursuit of physics is an activity intimately coupled with education. One conducts scientific research ideally to learn new things, and that enquiry is somehow incomplete until shared. Teaching science to others is in effect sharing the fruits of discovery made, not by oneself alone, but by some of the most creative people who have ever lived. It is not simply occupational parochialism that fosters my belief that science in general, and physics in particular, are much more than merely the source of better technology and higher lifestyles, but rather a precious intellectual legacy to pass on to future generations. And yet, as anyone knows who keeps abreast of the current state of education in America, Britain and elsewhere, science is one of the subjects least understood or favoured by the general public. The enormous divergence between the public perception of science and the profoundly interesting and important heritage that scientists know it to be, should be a matter of great concern. I have chosen to end this book, therefore, with an essay not on physics, but on the teaching of physics – or, more generally, on why science is worth knowing and how it might best be learned.

In matters of science and education it is to my wife, Dr Susan Brachwitz, and to my children, Chris and Jennifer, that I owe my greatest debt of gratitude. Besides the full-time occupations of university teaching and research, I have had, along with my wife, the responsibility and privilege of instructing our children from infancy onward in a home-based school. It was I, however, who received the most instruction, for what I have come to understand about the nature of learning and the stimulation of interest in science, I have learned with Susan's help from watching children develop in an atmosphere supportive of their natural instincts to explore and to discover.

In the practical matters of publishing, I would like to express my warm appreciation to Dr Simon Capelin, Senior Physics Editor at Cambridge University Press, for: initiating this project by requesting a book; responding enthusiastically when the book turned out to be not a textbook, but 'such an unusual proposal' instead; being patient while his author broke more deadlines than speed records in the completion of this work; and skilfully helping to transform ths submitted manuscript into the present volume of essays. This volume has also benefited significantly from the thorough reading and perceptive comments of my copy-editor, Beverley Lawrence.

I gratefully acknowledge as well the expert assistance of Madame I. Del Taglia in the preparation of the illustrations.

Finally, I would like to pay homage to a former mentor, Professor Francis M. Pipkin of Harvard University, whose capacity to do creative work in atomic, nuclear and elementary particle physics greatly impressed upon me during my student days how the horizons of a scientist need never be limited except by his own imagination and effort. Frank Pipkin was still involved in his extensive programme of research when he died in January 1992.

Mark P. Silverman

1

The unimaginably strange behaviour of free electrons

1.1 Variations on 'the only mystery'

Strangeness is a relative thing. With varying degrees of sophistication I have been thinking about physics for more than thirty years now, and this has no doubt both strongly and subtly influenced how the world presents itself to my eyes. There are laws and principles as familiar to me as the names of my children; most people are unaware of them and would not believe them even if informed.

As I leave my office, I ball up one last piece of scrap paper and toss it into the recycling box. The paper follows a nice parabolic arc as it lands. Were I to toss a rubber ball or a steel ball-bearing in exactly the same way, I know that (barring air resistance) it would follow the same path in the same time. All objects, irrespective of mass, chemical composition or any other physical property, fall at the same acceleration at the same location on the surface of the Earth. Galileo allegedly demonstrated this some four centuries ago. Yet surveys of science 'literacy' show that much of the American and British public readily subscribe to the Aristotelian notion that heavy objects fall faster than light ones.

It is dark out when I reach my car to start for home. The Moon lies suspended above one of the campus sports fields like an enormous orange. I know, however, that it is falling towards the Earth with an acceleration roughly 1/3600 that of the wadded paper I tossed some moments earlier. I have no fear of being crushed for, although it is falling, the Moon will never reach the Earth – not in my lifetime at least, if at all. An inward radial attraction, in fact, is what makes the Moon go around the Earth in a circular orbit. Again, most people would find that thought strange. Like René Descartes they imagine some force pushing the Moon tangentially around its orbit.

1

Upon reaching home I apply the brakes, and my car stops. If I did not apply the brakes, the car would eventually stop anyway (although not in a convenient location) because of friction. Excluding friction (and eventual obstacles), however, I know that the car would continue to move forward at constant speed forever. 'Move forward by itself forever?', I can hear one of my nonphysicist friends protest; 'Impossible! You have to push or pull an object to make it move'. And yet even now the Voyager satellites, long since detached from the rockets that launched them, continue to penetrate unimpeded the void of interstellar space.

The various consequences of the laws of gravity and motion addressed above *are* in some ways strange, but not unimaginably so. They are features of the macroscopic world to which physicists have reconciled themselves and which they can understand in terms visualisable to the mind's eye. Newton, for example, illustrated some three centuries ago in the *Principia* how a sequence of increasingly wide parabolic arcs of a free-falling projectile leads naturally to the circular trajectory of an orbiting satellite. Much later, during the second decade of the twentieth century, the mass independence of the law of free-fall found its explanation in Einstein's general theory of relativity which created the imagery of a conjoined space and time (space-time) warped by the presence of matter. The resulting contours of this incorporeal terrain constrain all matter to move along the shortest (actually, the extremal) paths or geodesics.

There is a qualitative difference between the tangible realm of classical physics, to which Newton's and Einstein's laws of motion and gravity belong, and the submicroscopic domain of the elementary particles and their composite structures. The principles governing the latter give rise to strange consequences that have never been – and most likely can never be – adequately interpreted in terms of objects or processes drawn from the world of macroscale experiences. The behaviour of such systems is unimaginably strange.

'A great physical theory ... when it is confirmed, takes on its own impersonal existence in the course of time, becomes completely detached from its originator, and is finally received as self-evident'[1]. So wrote the editor of a collection of Schrödinger's personal correspondence on wave mechanics. Having spent much of my professional life thinking about the intricacies of quantum physics, I am dubious that the theory will ever become self-evident (if, indeed, one can even

---

[1] *Letters on Wave Mechanics*, ed. K. Przibram (Philosophical Library, New York, 1967) vii.

characterise classical physics that way). Certainly, quantum mechanics is no longer the novelty that it was when its foundations were being laid in the 1920s, and a seemingly endless supply of basic textbooks makes this subject common knowledge throughout the physics community. Nevertheless, familiarity with the fundamentals of quantum mechanics has not by any means exhausted the surprises to which these principles still give rise.

The attribute of the quantum world that is responsible in large measure for its strangeness is that the denizens of this world, the elementary particles for example, propagate from one point to another as if they were waves, yet are always detected as discrete lumps of matter. There is no counterpart to this behaviour in the world we experience directly with our senses.

Imagine pouring a container of sand onto a flat plate with a small centrally located hole. A few centimetres below the bottom of the plate is a tiny movable detector that counts the number of sand particles arriving per unit of time. It would show, in accord with our expectations, that the greatest number of sand grains is registered directly under the hole; this number diminishes as the detector is moved transversely (i.e. parallel to the plate) away from the hole. Puncture another hole in the plate near the first one, and the sand pours through both in such a way that the total number of grains reaching the detector at any location is the sum of the number of grains reaching that point from each hole independently. In other words, opening up more holes can only *increase*, and never decrease, the total amount of sand reaching the detector per unit of time at any location.

But what if the experiment were performed with *electrons* rather than with sand? Quantum mechanics predicts that, if the aperture size and separation are comparable in magnitude to the wavelength of the electron (or, depending on the experimental configuration, some other characteristic length parameter), the scattered electrons, like light waves, should give rise beyond the perforated plate to an undulatory interference pattern (Figure 1.1). For a plate with two identical rectangular apertures the electron intensity, or number of electrons striking a unit area of the detector surface per unit of time, might be described mathematically as follows

$$I(\theta) = 2I_0(\sin^2 \beta/\beta^2)[1 + \cos 2\alpha]. \qquad (1.1a)$$

Here, the deviation of the electrons from the forward direction is measured by the angle θ; I_0 is the contribution to the electron intensity

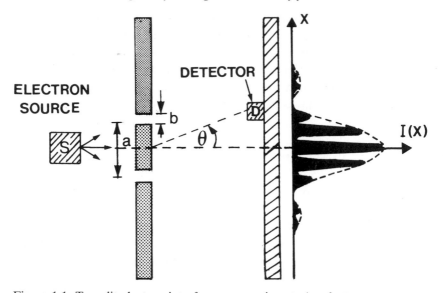

Figure 1.1. Two-slit electron interference experiment. An electron propagates like a wave from source S through the two slits to the detector D, but always registers as a discrete particle. The particle distribution $I(x)$ about the forward direction shows the oscillatory behaviour of wave interference. The broken line which envelops the interference fringes is the single-slit electron diffraction pattern.

from either aperture alone. The last factor in equation (1.1a) represents the *interference* between the components of the electron wave issuing from each aperture. The angle phase α upon which the interference pattern depends is given by

$$\alpha = (\pi a/\lambda)\sin\theta, \qquad (1.1b)$$

where a designates the distance between the centres of the apertures, and λ is the electron wavelength. The first factor in parentheses describes the *diffraction* of the electron wave through a single aperture, let us say of width b. For the sake of simplicity the aperture length is assumed to be much longer than the electron wavelength, in which case it will not significantly affect the passage of the electrons. The phase angle β of the diffraction pattern (which envelops the interference pattern) is expressible as

$$\beta = (\pi b/\lambda)\sin\theta. \qquad (1.1c)$$

The above expressions, which characterise the diffraction and interference of ideally mono-energetic electrons (or monochromatic light) by a particular configuration of slits, are provided only as an example. Nevertheless, they serve to highlight important differences between doing the experiment with sand grains or with objects that behave like waves. At the central maximum of the particle distribution, i.e. in the forward direction $\theta = 0$, the electron intensity can be four times – and not two times – the intensity from a single aperture. More generally, for a configuration of N slits, this intensity enhancement increases as N^2. At the locations of the interference minima, where β is an integral multiple of π, no electrons are detected. Thus, opening up a second aperture has resulted in fewer electrons arriving at certain locations than with only one aperture open. When the aperture width and separation are sufficiently larger than the wavelengths, however, one sees from the expressions for α and β that outside the forward direction the amplitude of the diffraction pattern becomes very small, and the fringes of the interference pattern become extremely narrow, eventually beyond the spatial resolution of the detector. The electrons, distributed nearly exclusively in the forward direction with a mean intensity of $2I_0$, would then seem to pass through the holes of the plate in much the same way as do grains of sand.

It is to be stressed that no electrons can be created or destroyed in this experiment; hence the total number of electrons received by the detector at all locations must be equal to the total number that has passed through all apertures. Somehow, scaling the properties of the particles and the width of the apertures from the macroscopic size of a sand grain down to the ultra-small size of an electron has radically altered the way in which the particles are distributed.

Richard Feynman, who had the good fortune to create his own version of quantum mechanics some twenty years after Schrödinger and Heisenberg developed theirs, characterised the wave-like interference of particles as

. . . a phenomenon which is impossible, *absolutely* impossible, to explain in any classical way, and which has in it the heart of quantum mechanics. In reality, it contains the *only* mystery[2].

The fact that electrons, which always register at a detector like hard little balls of sharply defined mass and charge, give rise in large numbers to an interference pattern may be surprising, but this is not the core of the mystery to which Feynman referred. The real enigma unresolvable

[2] R. P. Feynman, R. B. Leighton and M. Sands, *The Feynman Lectures on Physics*, vol. III (Addison-Wesley, Reading, 1965) 1-1.

by any mechanism of classical physics becomes apparent only when the electron flux (another word for intensity or number of particles 'flowing' through a unit area per unit of time) is reduced to such an extent that no more than one electron at a time passes through the apparatus. It is then that one must really come to grips with the implications of an electron wave.

The electron wave is *not* to be thought of as a water wave, sound wave or any other wave in matter which represents an actual physical displacement of a medium. Nor is it like a classical light wave composed of oscillating electric and magnetic fields which, though immaterial, is an expression of the classical electric and magnetic forces such a wave would exert on a unit electric charge. From the perspective of quantum mechanics classical waves are composed of enormous numbers of elementary quantum excitations. For example, 1 watt of pure red light of wavelength 650 nanometres represents an emission of about 3×10^{18} quanta of light (or photons) per second. The wave characterising the electrons is a *probability* wave; it allows one to calculate the probability of finding an electron within a given spatial region at a specified time.

Quantum mechanics does *not*, however, permit one to determine in which direction a *particular* electron (or any other elementary particle) that has been diffracted by some obstacle or aperture will eventually go. The arrivals of single electrons at the detector are random. Yet, according to theory, the random arrival of individual electrons in sufficiently large number should build up in time the same interference pattern that would be engendered quickly by a large electron flux.

In the course of my research I had often investigated the quantum behaviour of electrons theoretically. Although there is beauty and a measure of personal satisfaction in equations that reveal to the mind's eye striking new phenomena, the full implications of particle interference are so startling that they must be seen firsthand to be adequately appreciated.

In the mid-1980s, at a time when Japanese research laboratories were opening up to western scientists, I had the pleasure of being invited to the Hitachi Advanced Research Laboratory (ARL) at Kokubunji, a part of the Tokyo prefecture. Created only a short while before in the midst of the already flourishing Central Research Laboratory (CRL) devoted principally to applied research, the ARL was to be a sort of hybrid Japanese-style Bell Labs and Princeton Advanced Institute concerned with fundamental studies. I was there in part to help the electron holography group under the direction of Dr Akira Tonomura find novel ways to employ its craft.

The manufacture of electron microscopes is a speciality of the Hitachi Company, and central to the operations of the electron holography laboratory was a majestic state-of-the-art 150-kilovolt field-emission electron microscope. The source of electrons is a sharp tungsten cathode filament 100 ångströms wide (1 ångström $= 10^{-8}$ centimetres, about the size of an atom) from which electrons are drawn off by an electrostatic potential of a few thousand volts. One characteristic feature of the electron source deriving from the small tip size is the high degree of coherence of the electrons. 'Coherence' is a much-used word in physics, and even within the narrowed scope of electron microscopy has several connotations. It is effectively a measure of the extent to which electron interference can occur. For the present, suffice it to say that the electrons produced by field emission can give rise under appropriate circumstances to several thousand interference fringes – an order of magnitude improvement over other electron sources.

Shortly after my arrival at the ARL it occurred to me that, by employing a sufficiently attenuated beam, the electron holography group could make a film showing in real (or accelerated) time the evolution of the electron self-interference pattern, one electron at a time. Such a film, I suggested, would be of much use to physics teachers. Unfortunately, the project would not be possible, I was told, because the Hitachi chief management was still skittish over funding a purely basic research laboratory and looked particularly askance at experiments for 'classroom films'. Somewhere along the line, however, the management had a change of mind, for the experiment was eventually done, and a five-minute black-and-white video cassette was prepared which captured one of nature's most amazing phenomena[3]. (The promotional advantage of 'classroom films' as an aid to sales and recruitment did not go unnoticed for long. When I returned as visiting Chief Researcher to the ARL a few years later, Hitachi was in the midst of replacing the cassette with a full-length colour and sound film on electron interference.)

According to the de Broglie relation expressing one facet of the dual wave–particle behaviour of matter, the wavelength λ associated with electrons moving with linear momentum of magnitude p is given by

$$\lambda = h/p, \tag{1.2}$$

where h is Planck's constant ($\sim 6.6 \times 10^{-27}$ erg-second). (Another facet is expressed by the Einstein relation, $E = h\nu$, relating energy and frequency.) In the Hitachi experiment the wavelength of electrons emit-

---

[3] A. Tonomura, J. Endo, T. Matsuda and T. Kawasaki, Demonstration of Single-Electron Buildup of an Interference Pattern, *Am. J. Phys.* **57** (1989) 117.

ted from the field-emission tip and accelerated through a potential difference of 50 kilovolts was 5.4×10^{-10} centimetre (about one-tenth the Bohr radius of an unexcited hydrogen atom, and five orders of magnitude smaller than the wavelength of visible green light). With a kinetic energy of about 50 kilo-electron volts (or 50 keV)[4], the electrons moved relative to a stationary laboratory observer at a speed approximately one-half the speed of light ($c = 3 \times 10^{10}$ cm/s). Although fast by terrestrial standards, this speed v is still sufficiently below c that the Newtonian expressions for momentum ($p = mv$) and kinetic energy ($K = \frac{1}{2}mv^2$) lead to a value of λ reasonably close to that obtained from the exact relativistic expressions.

As an electron wave propagated through the barrel of the microscope (Figure 1.2) it was focussed by electromagnetic lenses and split by an electron biprism, a fine wire filament at a potential of about 10 volts placed between two parallel plates at ground potential. The biprism served in place of the two apertures. At the lower end of the microscope, single electrons impinged on a fluorescent film which emitted for each electron about 500 photons into a fibre plate that channelled the photons through to an underlying photocathode. Electrons ejected from the photocathode by the incident light were accelerated to 3 keV and entered a multichannel plate, a sort of honeycomb detector and electron multiplier by means of which the coordinates of the electron point image were determined with a position-sensing device. The arrival of an electron at a given channel was stored in an image processor, and the accumulating electron image could be viewed in real time on a TV monitor.

Initially there is no interference pattern; the detector simply registers, as expected, the random arrival of one electron at a time, each of which showed up on the monitor as a white dot. With the passage of time these random arrivals build up the classical two-slit interference pattern of fringes (Figure 1.3). The element of periodicity is barely discernible after the arrival of a few hundred electrons. It is definitely present, although not distinct, after a few thousand (squinting helps). After several tens of thousands of electrons the fringes stand out boldly; except for the wavelength scale they are indistinguishable from the fringes produced by light under comparable experimental conditions.

---

[4] One electron volt, or 1 eV, is the energy acquired by an electron falling through a potential difference of 1 volt. An energy on the order of 10 eV is required to ionise a hydrogen atom; this is roughly the energy scale at which chemical and biological processes occur. The energy equivalent of the electron mass (from Einstein's relation $E = mc^2$) is on the order of a million electron volts (MeV); this is roughly the threshold beyond which nuclear processes occur.

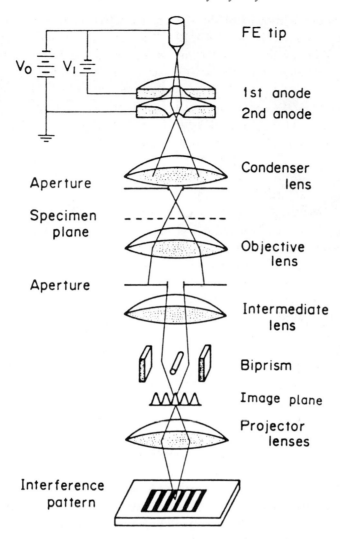

FE tip

V_0 V_1

1st anode
2nd anode

Aperture   Condenser
            lens

Specimen
plane

            Objective
            lens

Aperture

            Intermediate
            lens

            Biprism

            Image plane

            Projector
            lenses

Interference
pattern

Figure 1.2. Schematic diagram of an electron interference experiment with a field-emission (FE) electron microscope. An electron beam drawn by a high potential from the FE tip and accelerated through various focussing devices is split by the biprism and recombined in the image plane. The classically inexplicable outcome is that interference fringes are formed even when only one electron at a time passes through the microscope. (Courtesy of A. Tonomura, Hitachi Advanced Research Laboratory.)

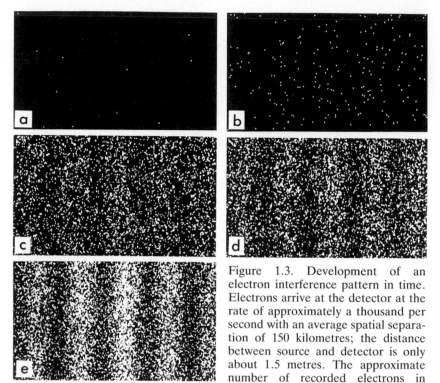

Figure 1.3. Development of an electron interference pattern in time. Electrons arrive at the detector at the rate of approximately a thousand per second with an average spatial separation of 150 kilometres; the distance between source and detector is only about 1.5 metres. The approximate number of recorded electrons in frames (a) to (e) are respectively 10, 100, 3000, 20 000 and 70 000. (Courtesy of A. Tonomura, Hitachi Advanced Research Laboratory.)

No one with a sense of curiosity could view this film and, knowing the circumstances, not be profoundly puzzled by the behaviour of matter on a subatomic scale.

How do the electrons 'know' where to go? Can there be some sort of cooperative effect between electrons emitted at different times that somehow leads to their preferential arrival at some locations and avoidance of others? This is highly unlikely, for the conditions were such that from a classical perspective one must regard the emissions as random, widely separated events. Let us consider a few relevant details.

Monochromatic waves, like frictionless surfaces and massless springs, are idealisations in physics; they cannot be produced by real sources that have been in operation for a finite length of time. The wave-like nature of the field-emission electrons is better represented by a wave packet. The length of the wave packet, designated the temporal coherence length l_t, is a rough measure of the spatial extent (along the direction of propagation) within which one is likely to find the electron. The electron is not emitted from the cathode tip at a precisely known instant in time; rather, there is an uncertainty (the coherence time t_c) in the duration of the emission. The coherence length is then

$$l_t = vt_c, \tag{1.3a}$$

where v is the mean electron speed. As a consequence of this temporal uncertainty, the energy of the electrons is also unsharp to an extent ΔE, where according to one expression of the uncertainty principle

$$t_c \sim \hbar/\Delta E, \tag{1.3b}$$

where $\hbar = h/2\pi$. From the de Broglie relation (1.2) and (nonrelativistic) expressions for the electron energy and momentum, one can derive the following formula for the coherence length

$$l_t = (2E/\Delta E)\lambda. \tag{1.3c}$$

In the Hitachi experiment the uncertainty in electron energy was about 0.5 eV. Estimated from relation (1.3c), the coherence length of the electrons was then on the order of a micron which, although small, is considerably greater than the electron wavelength and comparable in size to the diameter of the biprism filament. (One micron, or 1 μm, is 10^{-6} metre, about the size of some bacteria.) The wave packet comprised $l_t/\lambda = 2 \times 10^5$ electron wavelengths, and therefore represented a highly, although not perfectly, monochromatic electron beam.

How could one be sure that effectively only one electron at a time contributed to the interference pattern? By adjustment of the focal length of one of the lenses, the electron current reaching the detector was set to approximately 1000 electrons/second. Thus, one electron followed another at time intervals of about a millisecond. Moving at half the speed of light, the electrons were then separated from one another by about 150 000 metres, or a distance on the order of 100 000 times the length of the electron microscope! Under these circumstances an individual electron propagated from the source to the detector long before a succeeding electron was 'born'. So the question remains: How can a coherent macroscopic pattern be systematically created by randomly arriving noninteracting particles? It may be of interest to note that biologists face an analogous problem in accounting for patterns of animal coloration. How, for example, do the cells along the thin strip at the growing edge of a mollusc shell create intricate shell designs that are millions of times larger than the cells themselves[5]? The answer must lie

[5] C. Zimmer, Shell Game, *Discover* Vol. **13**, No. 5 (May 1992) 38–42.

in cell interaction – perhaps through diffusion of pigment activating chemicals. No such interaction can adequately explain electron interference.

The Hitachi experiment is not the first of its kind (although it was the first I had personally witnessed), but rather one of the last and most conclusive in a line of analogous experiments dating back to just a few years after Einstein proposed the existence of photons. In 1909, in an experiment remarkable for its technological simplicity, the British physicist G. I. Taylor[6] photographed the shadow of a needle illuminated by a light source so weak that on average only a few photons at a time impinged on the needle. After an exposure time of about 2000 hours the interference fringes of the diffracting pattern stood out as sharply as if a strong light source and much shorter exposure time had been employed. By contrast, the exposure time of the Hitachi experiment was about one hour.

The inadequacy of any explanation of interference phenomena based on the mutual interaction of electrons (or, as the case may be, photons) was noted by P. A. M. Dirac in his *Principles of Quantum Mechanics*, the Bible of quantum mechanics for several generations of physicists[7]. According to Dirac (page 9):

On the assumption that the intensity of a beam is connected with the probable number of photons in it, we should have half the total number of photons going into each component. If the two components are now made to interfere, we should require a photon in one component to be able to interfere with one in the other. Sometimes these two photons would have to annihilate one another and other times they would have to produce four photons. This would contradict the conservation of energy.

One might add that for electrons this would contradict the conservation of electric charge as well.

If there can be no cooperative effect between electrons, and if the presence in some spatial domain of an electron wave packet correspondingly implies the probability of finding an electron, it would seem that a given electron has to pass around both sides of the biprism wire simultaneously. Yet how can this be? The detector always registers an electron as an entire massive particle; one would need to explain how an electron could fragment and recombine. The Hitachi team did not

[6] G. I. Taylor, Interference Fringes with Feeble Light, *Proc. Cambridge Phil. Soc.* **15** (1909) 114.
[7] P. A. M. Dirac, *The Principles of Quantum Mechanics*, 4th Ed. (Oxford, London, 1958).

attempt to determine which path individual electrons took; had they done so, they would have found that an electron always passed to one side or the other, and never to both simultaneously. This act of looking, however, would have destroyed the interference pattern. The electron distribution would then no longer have been oscillatory, but rather the same as that of the grains of sand.

One cannot (as Feynman says) 'make the mystery go away by "explaining" how it works'. Nevertheless, there *is* a sort of explanation that stands as the central dogma of quantum mechanics; Dirac again expresses this clearly and succinctly in his *Principles* (page 9).

The new theory [i.e. quantum mechanics] which connects the wave function with probabilities for one photon, gets over the difficulty by making each photon go partly into each of the two components. *Each photon then interferes only with itself. Interference between two different photons never occurs.* [Italics added by the author.]

Dirac addressed himself to the interference of photons, but the principle applies without qualification to electrons as well.

The italicised phrase above is very important, indeed essential, to the standard interpretation of quantum mechanics. The self-interference of an electron, by which is meant the interference of the split electron wave packet, can occur only if the two components of the wave packet can overlap. Thus, qualitatively speaking, the difference in the 'optical path length' traversed by both components of the wave packet to a given point on the detector must not be much in excess of the coherence length l_t if self-interference is to occur.

Quantum theory furnishes the means to calculate the properties of the interference pattern produced by a beam of electrons, but it provides no means to envision the actual path of an electron. The very idea of a path or trajectory in a case where single-electron interference occurs is largely rendered useless by the uncertainty principle.

The mechanism, if one can even employ the word, of how an electron interferes with itself is indeed a mystery. But, Feynman notwithstanding, this is not the only mystery. It is just the beginning.

The self-interference of electrons is one manifestation of what is termed the wave–particle duality: the fact that 'particles' like electrons evince wave-like properties, and 'waves' like light evince particle-like properties. Examination of a diffraction or interference pattern does not reveal whether it has been made by electrons or by light (photons). This point is ordinarily deemed so obvious, once one accepts the wave–particle duality, that physics textbooks do not usually pursue it further.

Nevertheless, electrons and photons are quite different. Electrons have mass, $m_e = 9.11 \times 10^{-28}$ gram; photons (as far as is known) do not. (Examination of the spatial extent of the galactic magnetic field provides a photon mass upper limit of about 10^{-60} gram.) Electrons are electrically charged; photons are neutral. All photons carry one unit (in terms of $\hbar$) of intrinsic angular momentum. The intrinsic angular momentum, or spin, of the electron is $\frac{1}{2}\hbar$. This seemingly small difference in intrinsic angular momentum is the basis for major qualitative differences in physical behaviour. Photons are bosons, i.e. any number of them can be accommodated in a given quantum state; this, in essence, is the reason (from the standpoint of quantum mechanics) for the existence of classical light waves. Electrons are fermions, which signifies that at most only one electron can occupy a given quantum state; electrons cannot form classical waves.

Should not at least some of these properties – mass, charge, spin, statistics – affect the wave function and thereby distinguish electron from photon interference? They do – and at this point the behaviour of electrons becomes stranger still.

1.2 Electron interference in a space with holes

The development of classical physics documents in many ways the triumph of the field concept, Faraday's insightful vision of the transmission of forces between matter by means of an invisible, yet pervasive, medium. Of the various fields discerned by physicists those of gravity and electromagnetism are the most familiar and best understood. Gravity dominates the macroscale world of neutral matter, but is many orders of magnitude intrinsically weaker than electromagnetism. Two electrons an arbitrary distance apart repel one another with an electrostatic force some 10^{42} times stronger than their mutual gravitational attraction. Ordinarily (although not always, as we shall see later) gravity does not have a significant impact on the quantum behaviour of the elementary particles apart from those in highly collapsed, exotic systems like neutron stars, black holes and the like. Let us concentrate here on electromagnetism, and examine a quantum interference phenomenon arising from the existence of electric charge. Since light is electrically neutral, it is not expected to give rise to this effect.

All the phenomena of classical electromagnetism follow in essence from two sets of laws. On the one hand, there are Maxwell's equations, which describe the production of electric and magnetic fields from material sources of charge and electric current and from the spatio-temporal variation of the fields, themselves. Reciprocally, there is the Lorentz force law, which describes how the electromagnetic fields influence charged matter. Whether the Lorentz force is truly independent of Maxwell's equations is an interesting question, the answer to which depends essentially on what other assumptions one adopts about the properties of the fields. The point stressed here, however, is simply that (neglecting gravity) electrically charged particles interact with electric and magnetic fields; in the absence of such fields, classical physics provides *no* means by which the state of motion of charged particles can be perturbed. No E&M fields $\Rightarrow$ no E&M force!

This remark is important because within the framework of Maxwell's theory one customarily introduces, as a mathematical aid to the solution of problems, two auxiliary fields, the electromagnetic scalar and vector potentials, ϕ and A respectively. The electric and magnetic fields, designated E and B, can be expressed in terms of the spatial and temporal derivatives of the potentials as follows[8]:

$$E = -\operatorname{grad}\phi - \partial A/c\partial t, \tag{1.4a}$$

$$B = \operatorname{curl} A. \tag{1.4b}$$

One could, therefore, represent the Lorentz force law, which takes the form

$$F = e[E + (v/c) \times B] \tag{1.4c}$$

for a single particle of charge e moving with velocity v, in terms of these derivatives.

Nevertheless – and this again is essential to bear in mind – were it possible to envisage a region of space permeated *only* by the potentials yet *devoid* of all electric and magnetic fields, one would have to conclude from classical physics that a charged particle would experience no electromagnetic interaction in that region. For one thing, the electromagnetic potentials of a specified configuration of electromagnetic fields

---

[8] The gradient of a scalar function ϕ expressed in Cartesian coordinates is

$$\operatorname{grad}\phi = \partial\phi/\partial x + \partial\phi/\partial y + \partial\phi/\partial z.$$

Correspondingly, the curl (or rot) of a vector function $A = (A_x, A_y, A_z)$ is a vector with components of the form

$$(\operatorname{curl} A)_x = (\partial/\partial y)A_z - (\partial/\partial z)A_y,$$
$$(\operatorname{curl} A)_y = (\partial/\partial z)A_x - (\partial/\partial x)A_z,$$
$$(\operatorname{curl} A)_z = (\partial/\partial x)A_y - (\partial/\partial y)A_x.$$

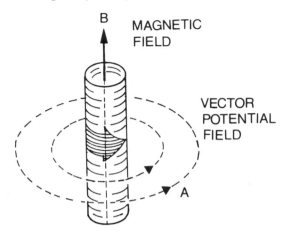

Figure 1.4. Static fields of an ideally infinitely long solenoid bearing a uniform current that circulates in the sense shown by the wide arrow. An axial magnetic field **B** fills the solenoid interior. A vector potential field **A** with cylindrical equipotential surfaces centred about the solenoid axis fills the space outside the solenoid (where the magnetic field is null).

are not unique; they can be changed in certain prescribed ways by a mathematical procedure known as a gauge transformation[9] without in any way altering the electromagnetic fields, Maxwell's equations and the Lorentz force law – and thus without changing the physical properties of the system. (A theory exhibiting this type of symmetry is said to be gauge invariant.) By contrast, real physical forces must be specified uniquely if classical physics is to lead to meaningful predictions.

The field configuration proposed above is not entirely a fanciful one. An infinitely long current-carrying wire wrapped tightly to form a cylindrical coil (or solenoid) of finite radius produces an axial magnetic field in the interior region with no return magnetic field in the exterior region (Figure 1.4). Nevertheless, the exterior region is permeated by a vector potential field with equipotential surfaces that form concentric cylinders about the solenoid. The sense of circulation of the vector potential and the direction of the interior magnetic field depend on the sense of current flow through the windings. Although nature does not provide physicists with infinite solenoids (any more than with frictionless bear-

---

[9] To effect a gauge transformation of a given set of electromagnetic potentials, ϕ and A, into a new set, ϕ' and A', select an appropriate gauge function Λ and calculate

$$\phi' = \phi - \partial\Lambda/c\partial t,$$

$$A' = A + \mathrm{grad}\,\Lambda,$$

where c is the speed of light.

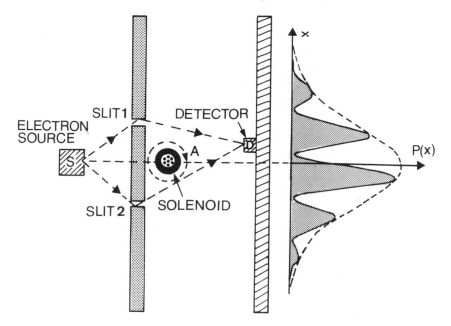

Figure 1.5. Schematic diagram of two-slit electron interference in the presence of a force-free vector potential field. The magnetic field inside the solenoid is directed perpendicularly into the page; the external vector potential field has a clockwise sense about the solenoid axis. Although the diffraction envelope remains undeviated from the forward direction, the interference fringes are displaced by a relative phase shift between the components of the electron wave issuing from slits 1 and 2. This phase shift is proportional to the magnetic field *within* the solenoid, the region from which the electrons are excluded.

ings), a real solenoid, to the extent that it is much longer than it is wide, can produce a magnetic field closely resembling the field of the ideal one. In any event, other geometrical configurations can be realised, and we will take up the practical details later.

What effect, if any, would such a solenoid have on charged particles if, to take a concrete example, it were placed midway between the two slits of the opaque partition employed in an idealised electron interference experiment (Figure 1.5)? The axis of the solenoid is oriented parallel to the plane of the partition (i.e. perpendicular to the page). It is to be understood that the solenoid is of sufficiently small diameter that it does not block the apertures, and that one should neglect the diffraction that would occur at the cylindrical surface irrespective of the presence of the electric current and associated internal magnetic field. It is further assumed that the space accessible to the electrons is limited to the solenoid exterior where the magnetic field is null and only the vector potential exists.

Clearly, in view of what was said above about classical electromagnetism, no influence on the electron interference pattern would be expected. In the quantum world, however, the concept of force is not as fundamental as the concept of potential. Potentials can influence the phase of an electron wave function to produce phenomena for which there are no classical analogues. Spatially varying potentials usually give rise to some kind of force, even when that force plays no direct role in the interpretation of a physical effect. For example, the gravitational potential of the Earth influences the wave function of a particle moving *horizontally* (near the Earth's surface) rather than falling vertically in response to the gravitational force. This leads to a physically observable quantum interference effect (to be discussed later) that depends on the acceleration of gravity, but which is *not* a direct consequence of the gravitational force, since the latter acts vertically downward only. In the present case, however, there is no electric or magnetic force at all in the region accessible to the electrons, and the ensuing effect is strange and thought-provoking even by the standards of quantum mechanics.

The first enunciation of what was eventually to become a major conceptual issue in the foundations of quantum mechanics was reported in 1949 by W. Ehrenberg and R. E. Siday[10] as something of an afterthought at the end of a long paper devoted to the correct determination of the refractive index in electron optics. The problem the authors addressed was of practical significance to the burgeoning field of electron microscopy where one needed to be able to determine electron trajectories through focussing devices. (The concept of an electron trajectory is meaningful when the wave-like nature of the electron is not involved; electron propagation can then be described in what is effectively a geometrical optics limit.)

Realising that the refractive index of an electron moving through the various focussing fields of an electron microscope 'contains the vector potential and not the magnetic field strength', Ehrenberg and Siday concluded: 'One might therefore expect wave-optical phenomena to arise which are due to the presence of a magnetic field but not due to the magnetic field itself, i.e. which arise whilst the rays are in field-free regions only'. To emphasise their point, the authors even described a hypothetical two-slit electron interference experiment not unlike that

[10] W. Ehrenberg and R. E. Siday, The Refractive Index in Electron Optics and the Principles of Dynamics, *Proc. Phys. Soc. (Lond.)* **B62** (1949) 8.

proposed above – but with the source of the magnetic field (e.g. a solenoid) left unspecified; they correspondingly deduced that, for each increment of 3.9×10^{-7} gauss-cm^2 in magnetic flux between the two slits, the electron interference pattern would shift by one fringe[11].

The physical quantity 'magnetic flux', which shall be represented here by Φ, should bring to mind the image of magnetic field lines 'flowing' through a surface. For a cylindrical region with constant axial magnetic field, such as the interior of the infinite solenoid, the flux is simply the product of the magnetic field strength B and the cross-sectional area (πR^2 for a solenoid of radius R). More generally, the magnetic flux of an arbitrary magnetic field through an arbitrary surface S is the areal integral

$$\Phi = \iint_S B \cdot dS, \qquad (1.5a)$$

where the dot or scalar product indicates that only the component of field perpendicular to the surface contributes.

As is well known from classical electromagnetism, the magnetic flux can also be deduced from the auxiliary vector potential field by means of a corresponding expression involving a contour or line integral

$$\Phi = \oint_C A \cdot dl \qquad (1.5b)$$

completely around the magnetic field lines (like a string around a bundle of straw). The entire closed contour C may well lie in a region, such as the external region of the solenoid, where the magnetic field, but not the vector potential field, is null. In a prescient remark concluding their paper, Ehrenberg and Siday commented: 'It is very curious that [there results] a phenomenon observable at least in principle with a flux; one expects a change in flux, but not steady flux, to have observable effects'. The 'change in flux' to which the authors referred recalls Faraday's law of induction and Maxwell's modification of Ampère's law whereby a time-varying magnetic or electric flux engenders, respectively, electric or magnetic forces. These are processes well within the purview of classical physics. Thus, Ehrenberg and Siday's discovery was indeed 'curious'. Unfortunately, as so often occurs in science, the novelty of a discovery is unrecognised by one's contemporaries and lies fallow until rediscovered under more propitious circumstances.

---

[11] The magnetic field strength of the Earth is on the order of 1 gauss near the Earth's surface. A small household bar magnet can be about 10^2–10^3 gauss.

The rediscovery took place independently ten years later. In a paper[12] regarded as a classic of quantum physics Y. Aharonov and D. Bohm discussed in all its puzzling detail the strange phenomenon that bears their names. The Aharonov–Bohm (or AB) effect has been controversial in one way or another for over three decades – an extraordinary situation for a science like physics. Within the community of physicists interested in such matters there are those (the majority) who believe the effect is an essential consequence of quantum mechanics, and that its observation has provided a fundamental confirmation of the theory. There are others who believe that the effect does not exist at all. And there are still others who, while admitting of the theoretical existence of the effect, are unconvinced that anyone has yet seen it. How can there possibly be such persistent divergence of opinion about the occurrence of a physical phenomenon?

At the core of the AB effect is the following characteristic of the electron wave function recognised by Dirac not long after the development of quantum mechanics. If $\psi_0(x,t)$ is the wave function of an electron at some point x and time t in a space free of electromagnetic potentials (and consequently of electromagnetic fields), then the wave function $\psi(x,t)$ of the electron in the presence of a time-independent vector potential field at the same space-time location can be expressed in the form

$$\psi(x,t) = \psi_0(x,t)\exp\{i(e/\hbar c)\int_{x_0}^{x} \mathbf{A}\cdot d\mathbf{l}\}. \qquad (1.6a)$$

The line integral in the phase of the wave function is taken along a path P, largely arbitrary, that connects the point of origin of the electron motion x_0 to the field point x. Since both the path and the mathematical form of the vector potential are arbitrary, the phase of the wave function is not uniquely prescribed. Nevertheless, the wave function of relation (1.6a) satisfies the quantum equation of motion (e.g. the Schrödinger equation for nonrelativistic electrons or the more general Dirac equation for relativistic electrons) when the vector potential is present, if ψ_0 is a solution when the vector potential is absent. The demonstration is quite straightforward, and, as far as I know, the above relation in itself scarcely raised any eyebrows *before* the AB paper pointed out unexpected physical consequences.

Note first of all that the indeterminate phase has no effect on

[12] Y. Aharonov and D. Bohm, Significance of Electromagnetic Potentials in the Quantum Theory, *Phys. Rev.* **115** (1959) 485.

measurements performed on an undivided electron beam, since the probability of finding an electron within some specified region, as well as the mean value of any physically observable property of the beam, depends on the expression $|\psi|^2 = \psi^* \times \psi$ from which the phase vanishes.

Consider, however, the two-slit interference experiment with the infinite solenoid. The electron wave function divides at the slits with the component issuing from slit 1 propagating around one side of the solenoid, and the component issuing from slit 2 propagating around the other side of the solenoid. After passage through the two slits and around the solenoid, therefore, the electron wave function comprises two terms

$$\psi(x,t) = \psi_1(x,t)\exp\{iS_1\} + \psi_2(x,t)\exp\{iS_2\}, \qquad (1.6b)$$

where ψ_1 and ψ_2 are the wave functions that would issue from slits 1 and 2 in the absence of the vector potential field, and S_1 and S_2 are the indeterminate phases of the form given in relation (1.6a). S_1 and S_2, however, are distinguished only by the 'path' taken by the electron. Actually, one can never know the path taken by the electron; all that really matters is that path P_1, to which phase S_1 is associated, lies on one side of the solenoid and path P_2, to which phase S_2 is associated, lies on the other.

It is not difficult to show that the phase *difference* between the two components of the wave function is then

$$S_1 - S_2 = (e/\hbar c) \oint_C \mathbf{A} \cdot d\mathbf{l} = e\Phi/\hbar c, \qquad (1.6c)$$

where C is a *closed* contour about the solenoid. The integral therefore represents the magnetic flux Φ through the solenoid interior. The phase difference, in contrast to the phase of each component, is *not* indeterminate, but rather an experimentally accessible quantity. From relations (1.6b) and (1.6c) it follows that the electron intensity at a distance x (from the forward direction) along the axis perpendicular to both the solenoid and the incident electron beam takes the form (for two identical slits)

$$I(x) = I(0)[1 + A\cos\{a(x) + 2\pi e\Phi/hc\}]. \qquad (1.6d)$$

Here A and $a(x)$ characterise the contrast and phase shift of the 'ordinary' two-slit interference pattern in the absence of the current-carrying solenoid. (The single-slit diffraction factor is not included in the above expression, since it is not relevant to the discussion at the moment.) The supplementary contribution to the phase which depends on magnetic flux is the AB effect. The flux-dependent term can be written as $2\pi(\Phi/\Phi_0)$, where the constant $\Phi_0 = hc/e = 3.9 \times 10^{-7}$ gauss-cm² is one 'fluxon', a fundamental unit of flux. If this expression really characterises the outcome of the proposed experiment, then, as Ehrenberg and Siday first predicted, a change in flux by one fluxon should shift the pattern by one fringe.

Within two years of publication of the AB paper, several laboratories reported experimental confirmations of the effect. Nevertheless, the AB effect was puzzling in almost every way; neither the theoretical existence, nor the experimental verification, nor the authors' interpretation of the effect, was readily accepted. In the words of Aharonov and Bohm[13]:

Although [our] point of view concerning potentials seems to be called for in the quantum theory of the electromagnetic field, it must be admitted that it is rather unfamiliar. Various of its aspects are often, therefore, not very clearly understood, and as a result, a great many objections have been raised against it . . .

The point of view of the authors, embodied in the title of their seminal first paper, is that the presumed auxiliary electromagnetic potentials, even though they are indeterminate, are in fact more fundamental than the electromagnetic fields. At least initially, before the deep significance of gauge invariance to field theory was widely recognised, this view rested largely on the notion of causality. To be consistent with commonly understood ideas of cause and effect implicit, for example, in the principle of special relativity, interactions in physics must be local; i.e. a particle can interact only with the fields in its immediate vicinity. This perspective is expressed in the very formulation of physical laws as differential equations. In the AB effect, however, the only field at the site of an electron is the vector potential field (or, in variations of the effect, the scalar potential) – and this field is indeterminate.

To many, however, the interpretation that electromagnetic potentials are more basic than electromagnetic fields was (and perhaps still is) difficult to accept. After all, although the electrons may be subject to the laws of quantum mechanics, the fields are still the classical fields of Maxwell's electrodynamics. There is nothing in the AB effect that requires a quantum theory of electrodynamics, and classical electro-

---

[13] Y. Aharonov and D. Bohm, Further Considerations on Electromagnetic Potentials in the Quantum Theory, *Phys. Rev.* **123** (1961) 1511.

dynamics can be formulated starting with either the fields or the potentials.

On the other hand, the alternative viewpoint, that the fields take precedence over (or are at least as fundamental as) the potentials, seemingly requires one to accept a most peculiar interpretation. Since the magnetic field is confined in a region of space inaccessible to the electrons, the particle–field interaction must occur *non*locally – i.e. by means of action at a distance. How can a magnetic field influence an electron that never passes through it?

One answer, maintained by a small minority, is that the whole issue is a tempest in a teapot: the AB effect does not exist except on paper. The argument to support this view is linked to the nonuniqueness of the vector and scalar potentials. It is possible to find a gauge transformation for which the new (i.e. transformed) vector potential field *vanishes identically* in the region outside the solenoid (or other current configuration). If this were indeed the case, the electrons could be made, by means of a purely mathematical manipulation, to pass through a region with neither a vector potential nor a magnetic field. Clearly one would not expect any influence on an electron in that case, and hence the effect predicted for a nonvanishing vector potential must be fictitious.

This reasoning, however, is not sound. The type of gauge transformation at issue not only removes the external vector potential, but effectively the *internal* magnetic field as well. It changes completely the physical system, and this is not permitted. Although there is wide latitude in the execution of gauge transformations, not every conceivable gauge transformation is an admissible one. A gauge transformation is a little like a change of coordinates; the selection of one coordinate system over another may afford more analytical convenience, but it must not change the physical system itself.

Ironically, the above point was already recognised in the 1949 paper of Ehrenberg and Siday who posed the question: 'One may ask if the anisotropy outside the [magnetic] field could not be avoided by an alternative value for *A* which also reproduces the field given . . .'. By 'anisotropy' the authors meant the presence of the vector potential in the theoretical expression for the electron refractive index. A short demonstration showed that this was *not* possible, and the authors concluded: 'It is readily seen that no vector potential which satisfies Stokes' theorem will remove the anisotropy of the whole space outside the [magnetic] field . . .'. Stokes' theorem – the key to resolving the gauge transformation 'paradox' – is the equality of relations (1.5a) and (1.5b). Expressed in words, the presence of a magnetic flux through a surface requires a nonvanishing vector potential field along some closed path bounding the surface. Any vector potential that does not satisfy Stokes' theorem for a specified magnetic field configuration is not acceptable. The Ehrenberg and Siday paper was apparently not widely read.

The AB effect, however, is a subtle one even for those who accept its existence. Indeed, what many physicists once thought (and perhaps still believe) the phenomenon to be is incorrect and violates basic physical principles! As depicted in the papers of both Ehrenberg and Siday and of Aharonov and Bohm, the phase shift engendered by the magnetic flux of the confined magnetic field redistributes the electron intensity out of the forward direction. So far, so good. Confusion arises, however, upon consideration of the actual manner of redistribution. In the words of Aharonov and Bohm, for example, the presence of the vector potential field of the solenoid has the following consequence: 'A corresponding shift will take place in the directions, and therefore the *momentum of the diffracted beam*'. [Italics added by the writer.] Is this in fact what the AB shift implies?

Recall at this point that, although the two-slit interference pattern has in principle an indefinite lateral extent, the fringe contrast falls off rapidly outside the central region of the single-slit diffraction 'envelope'. (See relation (1.1a) and Figure 1.1.) For the fringes to be visible, the transverse coherence of the electron beam must extend at least over the 'width' of the diffraction pattern (e.g. the region between the first two diffraction minima). This spatial coherence length l_s (to be distinguished from the longitudinal, or temporal, coherence length, l_t, introduced earlier) is given to good approximation by the relation

$$l_s = \lambda/2\delta, \tag{1.7}$$

where δ is the initial angle of divergence (i.e. angular deviation from the forward direction) of the beam at its source. As an example, the field-emission beam employed in the electron self-interference experiment of Section 1.1 had a divergence angle $\delta = 2 \times 10^{-8}$ radian and a wavelength of 0.054 ångström; thus, the transverse coherence length was $l_s = 1.4 \times 10^{-2}$ cm or 140 microns, two orders of magnitude larger than the temporal coherence length.

A common interpretation of the AB effect, expressed or implied in the expository literature, is that the shift in 'momentum of the diffracted beam' refers to the shift of the diffraction pattern. Indeed, Feynman, himself – one of the creators of quantum electrodynamics, the branch of physics that treats most comprehensively the interactions of particles and electromagnetic fields – had portrayed the AB effect as analogous to placing a strip of magnetic material (transparent to electrons) behind the partition with two slits; he showed that the resulting Lorentz force displaced the centre of the diffraction pattern (as, in fact, it would)[14]. This interpretation is *not* valid, however, for no such magnetic force is possible in a region ideally free of electric and magnetic fields. To represent the AB effect in this manner violates what is known as the Bohr correspondence principle.

Although quantum mechanics is more comprehensive than classical mechanics, there must be some means of relating both the quantum and classical descriptions of a system under conditions where the latter theory is also applicable. This is the correspondence principle, first enunciated and widely used by Niels Bohr in the years before a consistent and complete theory of quantum mechanics was formulated. The principle can be implemented in a variety of ways of which one of the most common is to consider the limiting case of a quantum expression as Planck's constant h approaches zero. As h vanishes, the laws of the quantum world and the classical world become one; the electrons stream through the apertures like (charged) grains of sand.

The quantum theoretical description represents this transition in the following way. First, the centre of the diffraction pattern falls at the location to which the extant forces displace the corresponding classical particles in accordance with Newton's laws. Second, the quantum interference pattern oscillates infinitely fast, so that no real detector could reveal the fringes. Thus, in the limit of vanishing h, the AB effect must vanish and the diffraction pattern of the electron beam must be undisplaced. What, then, is the physical consequence of the AB effect in the real world where h is not zero?

Careful analysis of the Aharonov–Bohm effect would show that the magnetic flux shifts the two-slit interference pattern asymmetrically *within* the single-slit diffraction pattern (Figure 1.5). But the centre of the diffraction pattern, itself, is *not* displaced in keeping with the condition that no classical electromagnetic force acts on the electrons[15].

The two-slit experimental configuration appears to be symmetric with respect to both slits. What determines whether the electron beam is

[14] R. P. Feynman, R. B. Leighton and M. Sands, *The Feynman Lectures on Physics*, vol. II (Addison-Wesley, Reading, 1965) Section 15-5.
[15] D. H. Kobe, Aharonov–Bohm Effect Revisited, *Ann. Phys. NY* **123** (1979) 381.

displaced laterally towards slit 1 or towards slit 2? It is the direction of the magnetic field, and consequently the sense of circulation of the vector potential, that breaks the symmetry. The field within the solenoid can be oriented either 'up' or 'down'; a change in the field orientation would reverse the direction of fringe shift, even though the electrons do not directly experience the magnetic field.

There is no mechanistic explanation. The AB effect, since it is discernible only in the interference pattern (which vanishes as h approaches zero) and not in the diffraction pattern (which remains unchanged from that of a force-free electron beam) is a uniquely quantum mechanical phenomenon and, as such, beyond the visual imagery of classical physics.

Theory aside, what does experiment have to say about the matter; has anyone actually observed the AB phase shift? Until quite recently, interpretation of the few electron interference experiments reporting AB-type phase shifts in the presence of structures designed to simulate an ideal solenoid was somewhat ambiguous. One such structure, for example, was a very fine magnetised iron 'whisker' less than a micron in diameter. Unfortunately, to ensure that no magnetic field lines permeate the region accessible to the electrons – whereupon critics could argue that changes in the electron interference pattern derive from the familiar Lorentz magnetic force – was rather difficult.

In a series of experiments extending through the 1980s, my Hitachi colleagues took up the challenge[16]. Through a happy marriage of basic science and advanced technology, the condition of a confined magnetic flux was produced by fabrication – *not* of solenoids – but of tiny (about 10–20 microns in diameter) toroidal (doughnut-shaped) permalloy ferromagnets. Unlike a real solenoid, where the currents in the windings produce an external magnetic field, the toroidal magnetic field lines form circular loops within the interior of the toroid (around the hole). Electron microscopy itself could be used as a check for magnetic field leakage, thereby permitting imperfect toroids to be discarded. And to make doubly sure the electrons would not penetrate the surface of the 200-ångström thick magnets, the Hitachi toroids were coated with a 3000-ångström thick layer of niobium and a copper layer also on the order of a thousand ångströms in depth.

To produce the AB effect, a shielded toroidal magnet was situated in the field-emission electron microscope above the electron biprism

---

[16] N. Osakabe *et al.*, Experimental Confirmation of the Aharonov–Bohm Effect Using a Toroidal Magnetic Field Confined by a Superconductor, *Phys. Rev.* A**34** (1986) 815.

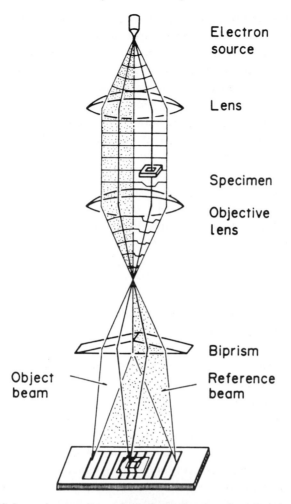

Figure 1.6. Schematic experimental configuration for observing the Aharonov–Bohm effect by means of electron holography. In the absence of the toroidal ferromagnet, the electron wave, split by the biprism and recombined in the image plane, generates straight, uniformly spaced reference fringes. With the toroid present, the portion of the electron wave diffracted by the toroid superposes with the reference wave at the image plane to generate a holographic image that preserves the phase (relative to the reference) of the electron wave function. (Courtesy of A. Tonomura, Hitachi Advanced Research Laboratory.)

(Figure 1.6). In the imagery of classical particles, an electron in the beam could pass either round the outside of the toroid or through the central hole. Although the magnetic field geometry differs from that of the ideal solenoid, there is still a net magnetic flux through any surface bounded by these two types of classical trajectories. In the imagery of classical waves, the portion of the electron wave that propagates round the outside of the toroid serves as a reference in an experimental configuration analogous to that of optical holography; this reference wave is split at the biprism and gives rise to a pattern of interference fringes upon recombination at the image plane (a photographic film). The component of the electron wave that diffracts through the central hole of the toroid, however, should incur an AB phase shift relative to the reference and thereby produce fringes shifted with respect to the reference fringes by an amount depending on the magnetic flux winding through the toroid. Since the electrons cannot penetrate the shielded toroid, the projection of the toroid onto the film ought to appear as a solid black annulus (flat doughnut). Were the toroid not shielded, one should see within the body of the annulus the continuity of the outside reference fringes and the displaced fringes in the hole.

The experiment was duly conducted with the results as just described (Figure 1.7). But alas – the sceptics were unmoved. The toroids, so it was claimed, were not perfect, or at any rate not close enough to perfection, and the spectre of the Lorentz force was again raised. Back to the drawing board (literally) went the Hitachi team.

To ensure beyond a reasonable doubt that the magnetic field of the toroid was adequately confined, the experimenters designed a low-temperature specimen stage to reduce the temperature of the toroid until the niobium layer becomes superconducting. A (Type I) superconductor displays the Meissner effect: upon transition to the superconducting state, it will suddenly expel a pre-existing magnetic field from its interior. However, given the geometry of the tiny shielded toroids, expulsion of the magnetic field from the outer niobium layer is tantamount to confining it within the inner permalloy magnet.

There was one small potential problem, however: the use of superconductivity entailed the discouraging possibility that no fringe shift might take place at all! A feature of superconductors, not unrelated to the Meissner effect, is that the magnetic flux penetrating a superconducting loop is constrained to half-integer values of the fluxon, hc/e. In other words, the AB phase shift in relation (1.6d) would take the form $2\pi(n/2) = n\pi$, where n can be 0, 1, 2, 3 ... etc. (One speaks of this as quantisation of magnetic flux, but this should not be construed to imply a quantisation of the magnetic field; the fields are entirely classical.) If, for some reason, the Hitachi toroids all produced even integer multiples of the fluxon, then the AB phase shift would be an integer

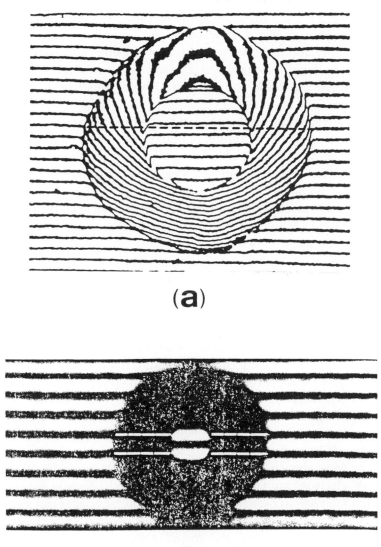

(a)

(b)

Figure 1.7. (a) Interferogram of an unshielded toroidal magnet showing the shift in fringes (i.e. lines of constant phase) for components of the electron wave passing outside the toroid or through the hole. Because the toroid is unshielded, electrons can also pass through the interior allowing one to see the continuity of the fringes. (b) A superconducting layer over the toroidal magnet shields the outside region from magnetic field leakage and prevents, as well, electron penetration. The interferogram shows a 180° phase shift between the two components of the electron wave that diffract around the toroid or through the hole. (Both courtesy of A. Tonomura, Hitachi Advanced Research Laboratory.)

multiple of 360°, and therefore not observable. On the other hand, if toroids could be produced for which the quantised flux turned out to be odd integer multiples of the fluxon, then phase shifts equivalent to 180° would occur thereby giving rise to complete fringe reversal between the space outside the toroid and the central hole. This would be clearly observable.

Fortunately, nature was not so perverse as to deny the researchers the fruit of their hard efforts. Toroids of both types were produced, and the expected phase shift was observed under conditions approximating more closely than ever before the ideal confinement of the magnetic field. The AB effect exists!

Or does it? Can one point again to deviations from ideality that vitiate the conclusion? More largely construed, the query addresses a basic aspect of how science works.

The Meissner effect, as is well known, does *not* exclude a magnetic field completely from the interior of a superconducting material. An 'evanescent' magnetic field (i.e. a field falling off exponentially with distance) can penetrate a superconductor to an extent (London penetration depth) that is ordinarily so small as to be negligible for a bulk substance. The penetration could, however, be significant for a thin film. The niobium layer covering the toroid is not a 'bulk' material, but is it sufficiently thick? From the thickness of the layers and the precision of their technique for measuring fringe shifts, the Hitachi group estimated that the leakage flux outside a toroid must be far less than 1/200 of a fluxon. According to quantum theory the ensuing Lorentz force should be negligible.

Well, what about electron penetration *into* the magnet? After all, 150 keV electrons are fairly energetic. Taking into account potentially relevant interactions between the electrons and the copper and niobium layers, the researchers estimated that about one out of every million electrons might penetrate the toroid sufficiently to experience a magnetic field. For a toroid 10 microns in diameter and an electron flux of about 10^{-5} ampere/cm^2, this amounts to some 50 electrons out of fifty million per second penetrating the toroid. Does this degree of imperfection invalidate the conclusion that the AB effect has been observed?

Certainly not. There is no such thing, in my opinion, as a 'definitive' experiment, an experiment so perfect or complete that it settles an issue for all time. There will always be deviations from ideal performance to which critics can point; there will always be new improvements which experimenters can make. The key concern, however, is whether or not the criticisms are valid and the improvements needed.

Within the framework of a mature science – indeed as a hallmark that a particular discipline is in fact a science – there must be objective ways for assessing the reliability of a given conclusion. It is insufficient as a mode of objection simply to list all manner of things that might conceivably be nonideal. To be taken seriously, a critic is obliged to demonstrate convincingly some relevant causal connection between the objection and the experimental outcome. In the present case the evidence, both theoretical and experimental, for the existence of the AB effect is consistent and reproducible. To sustain prior objections at this point one would need to explain how so striking a modification of the electron interference pattern could be produced by a miniscule trickle of penetrating electrons; or to demonstrate a flaw in the arguments leading to the presumed low level of penetrating electron flux; or to provide some alternative explanation (not involving the Lorentz force) of the observed phase shifts.

For most physicists concerned with the issue, the AB effect is theoretically real and experimentally confirmed. Many probably already believed in it, if only on the basis of theoretical self-consistency, well before the Hitachi experiments. Nevertheless, the scepticism that motivated these experiments was, I think, beneficial. The ensuing research not only convincingly demonstrated the AB effect, but led, as well, to novel instrumentation and new discoveries concerning quantised magnetic flux. Scepticism in science, if not also in the arena of everyday life, is a healthy attribute to the extent that it reflects an open-minded willingness to be convinced by new facts. That is how science advances.

What, finally, can one say about the interpretation of the AB effect regarding the fundamentality of electromagnetic fields or potentials? In quantum theory the electromagnetic potentials are no longer merely secondary fields that facilitate computation; they are needed at the outset in order that the theory be invariant under gauge transformations. Gauge invariance, once regarded as merely a curious feature of Maxwell's equations, has since been recognised as a most important symmetry to be maintained in all field theories. Indeed, this symmetry largely determines *a priori* the form of the interaction between particles and fields. Correspondingly, effects analogous to the AB effect are believed to occur, at least in principle, in areas nominally unrelated to electromagnetism such as gravity (general relativity theory) and the strong nuclear interactions (quantum chromodynamics). The interpretations of these effects are not always clear, nor are the experimental methods by which they might be observed. But that they are intrinsic to the theory and of fundamental significance is seemingly beyond doubt.

Recognition of the wider occurrence in physics of AB-like effects has led to reconsideration recently of the existence of an actual (i.e. electromagnetic) AB effect on light. Surprisingly, although such effects are not expected for completely neutral systems, the photon can nevertheless be influenced, at least in principle, by a magnetic flux. In classical electromagnetism, as a result of the linearity of Maxwell's equations, there is no mechanism by which light can interact with static electromagnetic fields or potentials (or with other light waves) in the absence of matter. Within the framework of relativistic quantum electrodynamics, however, a photon can, under appropriate circumstances, be transformed into an electron and positron pair, whose brief lifetime is so short that, to within limits provided by the uncertainty principle, no physical law is violated. During their emphemeral existence, these oppositely charged particles (rather than the original photon directly) can interact with a vector potential field to give rise, after their subsequent mutual annihilation back to another photon, to light scattering processes that depend on magnetic flux. The probability for the occurrence of such processes is extremely small, but that they exist at all highlights one of the seminal differences between classical and quantum electrodynamics.

The AB effect is a subtle one – at both the theoretical and experimental levels – and therein, in part, lies its great interest not only as a test of quantum mechanics and electrodynamics, but also as a stark reminder that physics is a human activity characterised by intellectual ferment, struggle and creativity. It is not simply a storehouse of equations, facts and procedures whose straightforward application instantly produces 'right' answers.

The confirmation of the AB effect does not, by any means, signify that the subject is an exhausted one. There are aspects to this phenomenon, as yet unexplored experimentally, that point to a physical reality stranger still than that revealed so far. One stands in awe at how devious and wonderful nature can be.

1.3 The two-electron quantum interference disappearing act

A loi..g grey wall separates the Hitachi grounds from the rest of Kokubunji City. The familiar orange Chuo ('Middle Central') Line passes close by one side of the wall taking commuters east to central Tokyo or west to outlying areas like the venerable city of Hachioji ('City of Eight Princes'). During my first visit to the Advanced Research Laboratory I lived in Hachioji on a high hill overlooking the city and affording a memorable view of Mt Fuji in the early morning hours when the air was clearest. Some years later I climbed Fuji with a Hitachi colleague, starting at twilight in a solemn torchlit procession of pilgrims, and caught a glimpse of the sunrise through a momentary parting of the thick curtain of mist that surrounded us. It was a moving experience. By then, I was living in Kokubunji within a short walk of the Laboratory.

Seen from outside the Hitachi wall, the grey tower of a company building looming up in the distance suggests just another industrial works. But inside, this mistaken impression evaporates before the extraordinary surroundings. Paths descend through wooded terrain to a large pond teeming with carp and lined with cherry and plum trees. Swans and ducks skim over the surface. A veritable botanic garden with a rich variety of trees, bushes and flowers identified by small placards surrounds the pond and adjacent smaller pools. Footbridges connect the mainland to a few small wooded islands upon which here and there a stone lantern or Japanese shrine nestles unobtrusively. Beyond the pond is an extensive sports field with outdoor amphitheatre.

These gardens and woods served me well. In the early afternoon and evening of most days I was wont to walk or jog around the pond, my head filled with thoughts of electrons, solenoids and quantum mechanics. The Aharonov–Bohm experiments – as indeed all quantum interference experiments to date with free electrons – were performed effectively one electron at a time. That is, even with the brightest sources available, the observed interference effects were all manifestations of single-particle self-interference. Dirac's dictum aside, the possibility of quantum interference with two or more electrons intrigued me, and I kept thinking about what types of effects could occur.

In the summer around the time of the o-bon festival – a holiday somewhat analogous in spirit, but not in celebration, to the European All-Saints' Day – the ARL and CRL staff held in the late afternoon a huge lawn party on the playing field. Then, when the sky had darkened sufficiently so that the first stars appeared, the crown jewel of the day's activities would begin: the *hanabi* or 'fire flowers', a spectacular display of fireworks lasting for almost an hour. I lay on my back in the soft grass watching burst after burst of brilliant particles and dreamily imagined them to be electrons shooting in all directions out of their source. What, I wondered, would two oppositely flying particles do if each encountered an AB solenoid at its own end of the sky?

The idea, although initially appealing, struck me after a few moments as uninteresting. Clearly, each particle would simply contribute to its own interference pattern of AB-shifted fringes. After all, once separated, the two particles go their merry way uninfluenced by one another. But I had not reckoned on what surely must be one of nature's strangest attributes, a quantum mystery no less profound than that of self-interference itself (Feynman's 'only mystery'). Some days later, when the inchoate images of celestial solenoids and *hanabi* electrons shooting through the heavens consolidated more soberly in my mind, I examined the problem systematically – and the results were surprising indeed.

To keep matters simple, imagine a compact source that produces wave packets with two electrons at a time. Like the fiery sparks of the *hanabi* there is no preferred direction for electron emission. One electron can fly out in any direction whatever, as long as the other electron emerges simultaneously in the opposite direction. How is one to create such an electron source? Well, I am not sure; this is, after all, a *Gedankenexperiment*. Perhaps one can fabricate a double-tipped field-emission cathode that emits pairs of coherent electrons, one electron emerging from each end. Perhaps there are atomic processes involving the correlated excitation and ionisation of two electrons. Or perhaps one can resort to the use of 'exotic' atoms, atoms containing elementary particles other than the familiar ones (electron, proton and neutron). If two electrons, for example, bind to a positive muon, there would result a muonic counterpart ($\mu^+ e^- e^-$) to the negative hydrogen ion H^-. A proton lasts forever (or at least many times the age of the universe); a muon, however, has a mean lifetime of about two microseconds after which it decays to other particles that flee rapidly from the scene of destruction. The decay of the muon in the $\mu^+ e^- e^-$ ion should leave two mutually repelling electrons. In any event, let us leave the technical details of a suitable electron source to the future.

Pick some point far to one side of the source (S) and place there an AB solenoid (with magnetic flux Φ_1) oriented perpendicular to the line between the source and the point; at a corresponding point on the

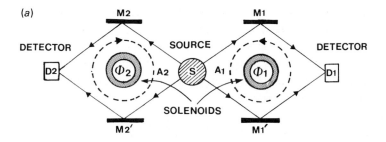

Figure 1.8(a). Schematic diagram of an electron interference experiment with pairs of correlated electrons. For each electron emitted by the source S in a particular direction, there is a corresponding electron emitted in the opposite direction. One does not know, however, which paths (source → mirror → detector) the electrons take through the right and left interferometers. The count rate observed at the detector of one interferometer depends on the magnetic flux in the other interferometer through which the counted electrons could not possibly have passed.

opposite side of the source locate a second AB solenoid (containing flux Φ_2) with its axis parallel to the first (Figure 1.8a). Let us suppose the current through the windings of the two solenoids circulate in the opposite sense so that the internal magnetic fields of the solenoid are antiparallel; it does not really matter – the principal results are not qualitatively changed.

Now locate four electron 'mirrors', as shown in the figure. If an electron – let us call it electron 1 – heads in just the right direction to reflect from mirror M1 into a detector D1, then the companion electron – electron 2 – will reflect from mirror M2′ into detector D2. Correspondingly, if electron 1 reflects from mirror M1′ into detector D1, then electron 2 reflects from mirror M2 into detector D2. The distance between source and mirror is the same for all four mirrors; likewise the four (shortest) mirror–detector separations are all equal. In this way there is no phase shift in the electron wave function arising from a difference in the electron optical path length. The two classically imagined paths of electrons 1 and 2 form complete loops about the respective solenoids S1 and S2, and it would seem then that all conditions for an AB interference effect at each detector are met.

It is to be assumed that the two solenoids (S1 and S2), and correspondingly the two detectors (D1 and D2), are far apart from one another. So far apart that the observers may not even be able to see or communicate with one another. Perhaps the observer at D1 is not even aware that there is someone to switch on D2 and observe the other electron. From the perspective of observer 1, therefore, electrons are simply arriving regularly and being counted at D1. 'Why should it matter if other electrons are spewing out elsewhere?', he might ask. If an electron takes the path S–M1–D1, the wave function of the electron incurs the (nonunique) phase shift α_1 where, in accordance with relation (1.6a) of the previous section

$$\alpha_1 = (e/\hbar c) \int_{\text{Path S–M1–D1}} A_1 \cdot dl. \tag{1.8a}$$

Similarly, if the electron takes the path S–M1'–D1, there results a phase shift α_1'

$$\alpha_1' = (e/\hbar c) \int_{\text{Path S–M1'–D1}} A_1 \cdot dl. \tag{1.8b}$$

Corresponding phase shifts α_2 and α_2' involving the vector potential field A_2 of solenoid S2 are incurred by the electron wave that propagates from the source S to D2 via the mirrors, M2 and M2', respectively. Observer 1 does not concern himself with the effect of vector potential A_2 in his vicinity because the electrons reaching detector D1 do not make a circuit around solenoid S2.

What does observer 1 predict will be the outcome? To the extent that he ignores the electrons emitted towards observer 2 – and, for all he knows, may not even be counted – the first observer might reason as follows. The wave function – or at least the only part of it relevant to his own experiment – should be the sum of two probability amplitudes, one for each path the electron could take to detector D1. Thus, to within a constant factor, the net amplitude for arrival of an electron at D1 is

$$\psi(\text{D1}) \sim \exp(i\alpha_1) + \exp(i\alpha_1'). \tag{1.9a}$$

Observer 1 would then deduce that the probability of an electron being received at D1 is proportional to $|\psi(\text{D1})|^2$ or

$$P(\text{D1}) = (1/2)[1 + \cos(e\Phi_1/\hbar c)], \tag{1.9b}$$

which is in effect a special case of relation (1.6d) of the previous section. The normalisation factor '1/2' assures that the maximum probability is unity or 100%. We have also made use of Stokes' law, discussed earlier, which in the present case requires that

$$\alpha_1 - \alpha_1' = e\Phi_1/\hbar c, \tag{1.9c}$$

$$\alpha_2 - \alpha_2' = e\Phi_2/\hbar c. \tag{1.9d}$$

Observer 2, reasoning in a similar way that his experiment is independent of that of the distant observer 1, would deduce an analogous expression involving flux Φ_2. Together, the two observers would infer the following joint probability for an electron to be received at both D1 and D2

$$P(D1,D2) = P(D1)P(D2)$$
$$= (1/4)[1+\cos(e\Phi_1/\hbar c)][1+\cos(e\Phi_2/\hbar c)]. \quad (1.9e)$$

The results may seem satisfying, for the probability inferred by each observer depends only on the magnetic flux of the solenoid in 'his' part of the universe. The only problem is that the predicted outcomes (relations (1.9b) and (1.9e)) and the whole mode of thinking are *incorrect*.

The two electrons are not emitted independently since their 'paths' (actually their linear momenta) are correlated; according to quantum theory this correlation persists, no matter how far apart the electrons travel. If the observer at D1 determined the path of arrival of an incoming electron, he would know without having to make a measurement (if he were aware of the correlated emission) the path taken by the other electron. Of course, if he *did* determine the electron path, there would no longer be any quantum interference. Nevertheless, according to the hypothetical conditions of the *Gedankenexperiment*, an electron necessarily arrives at D2 by a specified (albeit unknown) path if an electron arrives at D1, and one must determine the joint probability of electron detection at the outset.

The amplitude that one electron arrives at D1 via mirror M1 and therefore the other arrives at D2 via mirror M2' is proportional to the product of the phase factors for each route

$$A(M1,M2') \sim \exp(i\alpha_1)\exp(i\alpha_2'). \quad (1.10a)$$

Likewise, the amplitude that one electron arrives at D1 via mirror M1' and therefore the other arrives at D2 via mirror M2 is

$$A(M1',M2) \sim \exp(i\alpha_1')\exp(i\alpha_2). \quad (1.10b)$$

The wave function $\psi(D1,D2)$ of the detected two-electron system is proportional to the sum of the above two amplitudes, and the joint probability of electron detection, given by $|\psi(D1,D2)|^2$ now takes the form

$$P(D1,D2) = (1/2)[1+\cos\{(e/\hbar c)(\Phi_1 - \Phi_2)\}]. \quad (1.10c)$$

From the perspective of classical physics the above hypothetical experiment with correlated (as opposed to independent) electrons poses a curious dilemma in several ways. First, until one gets used to it, the idea of the AB effect, itself, where electrons are affected by passing *around*, and not through, a magnetic field is rather curious. But in the present configuration, the signal detected by one observer *also* depends on the distant magnetic field around which 'his' detected electrons have *not* passed. Or, phrased differently, it depends on the electrons that go to the *other* observer even though the latter can be arbitrarily far away! Suppose, for example, the magnetic flux is the same within the two solenoids; irrespective of the magnitude of the flux, the electrons, according to the correct relation (1.10c), would arrive at both detectors with 100% probability. By contrast, if the flux through S1 had been set to produce a phase shift $e\Phi/\hbar c$ of 180°, then observer 1, in the erroneous belief that his experiment was independent of that of the other observer, would have deduced from relation (1.9b) that the probability of electron arrival at D1 is *zero*.

There is nothing in itself strange about two particles being correlated and flying off in opposite directions; this could occur as well in classical mechanics if an initially stationary object exploded into two pieces. But once the pieces are separated, the motion of one would not be expected to influence, or be influenced by, subsequent measurements made on the other. What has happened to the 'localness' of physical interactions in quantum mechanics?

Perhaps the reader is thinking that the nonlocality manifested by the above two-electron AB *Gedankenexperiment* is an artificial product of the experimental condition whereby, if observer 1 receives an electron, then he knows for certain that observer 2 has received an electron. In other words, the experimental configuration is such that one cannot calculate the signal at one detector without it being a joint detection probability for both detectors (relation (1.10c)).

Let us modify the experiment, therefore, so that when observer 1 receives an electron, he *will not know* to which detector the other electron has gone. At the former location of detectors D1 and D2 put two beam splitters, BS1 and BS2, that divide the intensity of an incoming electron beam equally (Figure 1.8b). Thus, an electron incident on BS1 from mirror M1 or M1' has a 50% chance of reflecting from the surface and a 50% chance of being transmitted. Likewise for electrons incident on BS2 from mirrors M2 and M2'. Now, instead of having only two detectors as before, let us place *four* detectors (D1, D1', D2, D2'), one on each side of each beam splitter. With these modifications not

(b)

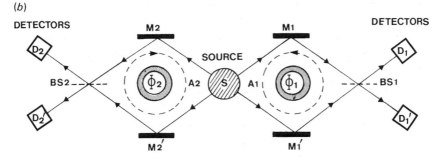

Figure 1.8(b). Diagram of correlated electron interference experiment with addition of beam splitters BS1 and BS2 to the configuration of Figure 1.8(a) (and corresponding addition of two more detectors). The probability for detecting an electron at a specified detector, irrespective of the fate of the paired electron, is entirely independent of magnetic flux and, in fact, shows no quantum interference effect at all. Interference occurs only in the joint detection of two electrons.

every electron incident on the mirrors M1 or M1′ will necessarily go to the detector D1 of observer 1; nor can observer 1 know if the second electron of the correlated pair emitted by the source is received at D2 or D2′.

Alas, the modified two-electron AB experiment does *not* lead to results more compatible with our classical conceptions of locality. If anything, the outcome is stranger than before.

One can readily determine, by extension of the foregoing reasoning leading to the amplitudes (1.10a) and (1.10b), the probability amplitude for each potential pathway of an electron wave from source to mirror to beam splitter to detector[17]. From these amplitudes follows the joint probability for receiving two electrons at any two detectors. For example, the joint probability that one electron arrives at detector D1 and the other electron at detector D2 is

$$P(D1,D2) = (1/4)[1 - \cos\{(e/\hbar c)(\Phi_1 - \Phi_2)\}] \qquad (1.11a)$$

while the joint probability that one electron arrives at detector D1 and the other electron at detector D2′ is

$$P(D1,D2') = (1/4)[1 + \cos\{(e/\hbar c)(\Phi_1 - \Phi_2)\}]. \qquad (1.11b)$$

---

[17] The amplitude for electron transmission through a (nonabsorbing) beam splitter has a phase shift of 90° relative to the amplitude for reflection. This property, which also characterises the reflection and transmission of light under comparable circumstances, derives from particle and energy conservation.

These expressions are similar to relation (1.10c) characterising the first thought experiment.

However, suppose observer 1 is not interested in where electron 2 goes. In fact, suppose that detectors D2 and D2′ are not even turned on, and that *nobody* knows what has happened to electron 2. Observer 1, nonetheless, sits by detector D1 and assiduously notes down the electron counts. What signal does he receive?

The probability that an electron goes to D1 *irrespective* of the detector to which electron 2 goes is simply

$$P(D1) = P(D1,D2) + P(D1,D2') = \tfrac{1}{2}. \qquad (1.11c)$$

A constant! The signal that observer 1 receives with his detector alone is completely independent of the magnetic flux of either solenoid. The AB effect seems to have completely disappeared.

In fact, *all* quantum interference has disappeared. Had the geometrical path lengths for the pathways S–M1–D1 and S–M1′–D1 not been equal (as initially specified), then – quite apart from the presence or absence of magnetic flux through any solenoid – there would occur a relative phase shift between the electron waves taking one or the other of these pathways. One would then expect the usual two-slit type of quantum interference to occur. But it does not occur. The phase shift engendered by an optical path length difference would appear in the argument of the interference term (the cosine function) of both relations (1.11a) and (1.11b). Since these two interference terms have opposite signs, they would again vanish when summed to give $P(D1)$, the probability that observer 1 receives an electron.

The *joint* probabilities, $P(D1,D2)$ and $P(D1,D2')$ *do* show an AB effect. Thus, when either the observer at D2 or the one at D2′ correlates his electron count rate with observer 1, the latter becomes aware of an AB effect. But, if neither the observer at D2 nor the one at D2′ bothers to participate in the experiment, then observer 1 detects no quantum interference effects at all. How can this be? How can observers at one end of the universe destroy the quantum interference of electrons at the other end simply by deciding *not* to observe? Surely this is most odd.

As with other uniquely quantum phenomena, the above *Gedankenexperiment* – which some day will no doubt be performed as a real experiment – has no explanation within classical physics. We cannot satisfactorily account for these results through any imaginable behaviour of particles and waves such as we find in the macroscale world.

It should be brought out explicitly at this point that – despite the presence of a vector potential field in the region of space accessible to the electrons – the AB effect even as originally described (with diffraction of a single-electron wave function around one solenoid) is, itself, an intrinsically nonlocal phenomenon. The presence of a vector potential field does not *per se* make the AB effect an *effect* – i.e. something observable – unless the pathways potentially available to the charged particle circumscribe the confined magnetic field. In this sense the AB effect reflects the topology (global or nonlocal geometrical features) of the multiply connected space through which the particles propagate.

But quite apart from anything specific to the AB effect, quantum mechanics manifests an essential nonlocality deriving from the wave description of matter. Once linked together in a single quantum system, quantum particles remain a single system even after they have separated sufficiently far that there can be no information exchange or physical interaction between them capable of affecting an experimental outcome. These results *are* strange, even within the framework of quantum mechanics, because most of our past experience with quantum interference phenomena concerned principally the self-interference of single particles. The hypothetical system described above, however, involves inextricably 'entangled' quantum states of two particles (to employ a terminology first introduced by Schrödinger). No matter how far apart the two particles separate, they constitute a single quantum system manifesting the bizarre effects of nonlocality.

This may seem absurd from the perspective of the familiar experiences which define for most of us the nature of 'physical reality'. If so, we are in good company, for Einstein himself was sorely plagued by these strange implications of quantum theory.

In 1935, in a paper[18] that has subsequently become a wellspring of voluminous discussions and experimental tests of quantum mechanics, Einstein and his collaborators Boris Podolsky and Nathan Rosen raised the issue known today as the EPR paradox. Does quantum mechanics provide a complete description of physical reality? EPR posed the question and answered it negatively. What, after all, *is* physical reality? According to EPR:

If, without in any way disturbing a system, we can predict with certainty (i.e. with probability equal to unity) the value of a physical quantity, then there exists an element of physical reality corresponding to this physical quantity.

[18] A. Einstein, B. Podolsky and N. Rosen, Can Quantum-Mechanical Description of Reality Be Considered Complete?, *Phys. Rev.* **47** (1935) 777.

And, insisted EPR,

... every element of the physical reality must have a counterpart in the physical theory

if a theory is to be regarded as 'complete'.

How well does quantum mechanics do when judged by these criteria? Not very well, it seems. EPR provided their own example for disqualifying quantum mechanics as a complete theory, but our first two-electron AB experiment will also do just as well. As stated before, without in any way disturbing the electron emitted to the left, observer 1 can determine which path it takes to detector D2 by determining which path the electron emitted to the right follows to detector D1. By the EPR criteria, then, these electron paths are elements of physical reality and must have a counterpart in the physical theory. But we know that, if the ensuing quantum interference is not to be destroyed, quantum theory does *not* permit us to know which path either electron has taken. To know this information is equivalent to knowing the transverse components of coordinate and momentum of each electron to a degree of precision higher than that permitted by the uncertainty principle. In quantum theory, these various paths available to the electrons are not elements of an objective physical reality as much as they are elements of a potential reality.

This does not mean, of course, that quantum mechanics is necessarily an incomplete theory. Rather, the EPR definition of reality may not be adequate, an objection that they, themselves, anticipated:

One could object to [our] conclusion on the grounds that our criterion of reality is not sufficiently restrictive. Indeed, one would not arrive at our conclusion if one insisted that two or more physical quantities can be regarded as simultaneous elements of reality *only when they can be simultaneously measured or predicted.* ... This makes the reality of [such quantities] depend upon the process of measurement. ... No reasonable definition of reality could be expected to permit this.

Einstein, Podolsky and Rosen's assertion notwithstanding, physical reality is what it is: the strange reality depicted by the quantum theory. Does an alternative description of physical reality exist? Einstein believed that such a description would some day be found. But physicists have searched for decades, and are searching still; no alternative theoretical framework has been created, as far as I am aware, that is as successful as quantum theory.

By the time Einstein advanced the views expressed in the EPR paper, he was already largely regarded as out of the mainstream of modern physics. The paper met with a barrage of rebuttals and criticism although, as Einstein wryly noted, no two critics objected to the same thing. Einstein died four years before the article by Aharonov and Bohm appeared, and it is unlikely, I would surmise, that he ever saw the paper by Ehrenberg and Siday. I have often wondered what Einstein would have said about the AB effect, which so alters our conception of physical reality not only in the domain of mechanics, but in electro-magnetism as well. As one interested in the foundations of electro-dynamics throughout his life, would he have considered the primacy of potentials over electric and magnetic fields a violation of his cherished beliefs, or would he have said – as the young Einstein rashly did upon hearing of Bohr's theory of light production in 1913 – 'The theory ... must be right'[19]? Would he, the master geometer of physics, have been pleasantly surprised or appalled at a quantum phenomenon dependent on the topology of space? We will never know.

1.4 Heretical correlations

In addition to electrical charge, the electron is also endowed with an intrinsic angular momentum, or spin, of $\frac{1}{2}\hbar$. Whereas charge is respon-sible for the interactions leading to the Aharonov–Bohm effect, spin gives rise, at least indirectly, to different quantum interference effects that are in some ways even more remote from our classical expectations. Although the AB effect cannot be accounted for in terms of electric or magnetic forces, it is nevertheless a consequence (albeit a quantum consequence) of classical electromagnetic fields. However, there are consequences of electron spin that have *no* classical roots whatever. Unlike the (single-electron) AB effect, the electron phenomena to be considered here have not yet been observed in the laboratory, but they must exist if our current understanding of the quantum behaviour of matter is correct. These effects are not only impossible to reconcile with the imagery of classical physics, but appear to challenge, as well, the traditional interpretation of particle interference, i.e. Dirac's dictum: a particle can interfere only with itself.

It has long been a fundamental tenet of wave theory – pertinent as well to the 'wave mechanics' of matter – that wave *amplitudes*, and never intensities (i.e. the squares of amplitudes), interfere. One might

---

[19] Cited in W. Moore, *Schrödinger* (Cambridge University Press, Cambridge, New York, 1989) 137.

well imagine, therefore, that the development of an interferometer, in which the superposition of separate light intensities produced an interference pattern, would be viewed with considerable scepticism. And indeed that was exactly the response of many physicists to the intensity interferometer of R. Hanbury Brown and R. Q. Twiss (HBT)[20].

Developed in the 1950s for the purpose of measuring stellar diameters by a method less sensitive to mechanical vibrations or atmospheric distortions than suspending a Michelson interferometer[21] at the end of an optical telescope, the HBT instrument functioned as follows (Figure 1.9). Light from a star was received at two spatially separated photodetectors whose electrical outputs were passed through 'low-pass' filters. The filters suppressed components of the electric current oscillating at frequencies outside the domain of the radiowave spectrum, i.e. outside the range of about 1–100 megahertz (1 MHz = 10^6 oscillations per second). The two filtered currents were then multiplied together electrically and averaged over a prescribed time interval. The resulting number, a measure of what is termed the cross-correlation of the incident light, produced an oscillatory curve when plotted as a function of the separation of the two detectors – a clear sign that some kind of interference had occurred.

What is most noteworthy here is that the photodetectors are so-called square-law devices: the electrical output is proportional to the incident light power flux (i.e. intensity). It is the multiplication or correlation of two intensities (not amplitudes) that has produced an interference pattern. How can two intensities interfere?

Actually, the phenomenon does no violence to any known physical principle and can be explained quite simply within the framework of classical wave theory in a way that an electrical engineer would find satisfying and far from surprising. Briefly, each point on the stellar surface gives rise to broad wave fronts at the Earth that illuminate *both* photodetectors of the interferometer. At the surface of each detector, therefore, there occurs a superposition of numerous amplitudes emitted from different locations on the star at different frequencies and with random time-varying phases. What remains from the filtering of the

---

[20] R. Hanbury Brown and R. Q. Twiss, A New Type of Interferometer for Use in Radioastronomy, *Phil. Mag.* **45** (1954) 663.

[21] Conceptually, Michelson's stellar interferometer is another version of the classic two-slit experiment. In lieu of apertures, there are two mirrors a distance *d* apart, each one reflecting an image of the light source (a star). The two images are made to superpose on a third mirror thereby giving rise to a pattern of interference fringes similar to that of Figure 1.1.

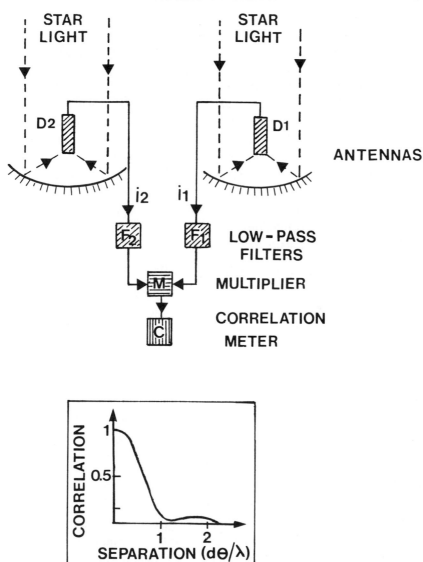

Figure 1.9. Schematic diagram of the Hanbury Brown–Twiss intensity interferometer. Light from an extended source is received by two photodetectors sensitive to the light intensity. The output current of each detector is passed through
a filter that admits only components oscillating at low frequencies (corresponding to beat frequencies among the incident optical waves). The two filtered
outputs are multiplied electronically, averaged over time, and recorded by the
'correlation meter' as a function of receiver separation. Although light intensities are not expected to interfere, the resulting oscillatory correlation curve
reveals that some kind of interference has occurred.

electrical signals are low frequency 'beats' produced by the interference of waves at neighbouring optical frequencies. For each pair of light-emitting points on the star the phase associated with a given beat frequency at a given detector consists of two parts: (a) a well-defined component determined by the optical path lengths between the point sources and the particular detector, and (b) a random time-varying component resulting from the *difference* in initial random phases of the superposing waves. However – and this is the key point – at a given instant the random phase difference associated with a particular beat frequency is the *same* for *both* detectors, since (in the imagery of classi-cal optics) the same broad wave fronts sweep over both detectors. Thus, multiplication and time-averaging of the filtered signals do not lead to the vanishing of all correlations, but – for each pair of interfering waves – to the correlation function $c(d)$,

$$c(d) \sim I_1 I_2 \cos(2\pi d\theta/\lambda) \qquad (1.12a)$$

proportional to the mean light intensity (I_1, I_2) at each detector. The phase of the correlation function varies with detector separation d, angular separation (as seen from the Earth) of the point radiators θ, and mean wavelength λ. The above result must then be averaged over all pairs of points on the stellar surface and over all contributing optical frequencies in order to obtain the net signal. Surprisingly, the result is quite simple; $c(d)$, suitably normalised, is effectively the square of the fringe visibility produced by the same light in the Michelson stellar interferometer with mirror separation d. The visibility, or contrast, of the fringe pattern

$$V(d) = \frac{I_{max} - I_{min}}{I_{max} + I_{min}} \qquad (1.12b)$$

is defined as the difference in intensity between neighbouring points (near the centre of the pattern) of maximum and minimum brightness divided by the sum of these intensities; as a function of mirror separa-tion, $V(d)$ can range between values of zero and unity[22].

As one may have expected, it is ultimately the amplitudes, and not really the intensities, that interfere in an intensity interferometer. Nevertheless, the correlation depends on the product of intensities,

---

[22] When the mirrors are close together, the superposed amplitudes have a well-defined relative phase; the intensity of the dark fringes approaches zero, and the visibility is close to unity. When the mirrors are far apart, the two images constitute essentially independent sources; there is practically no interference, and the visibility vanishes.

rather than the product of wave amplitudes, in marked contrast to our archetypical example of two-slit interference discussed previously. Also, since the highest frequency of the detected beats ($\sim$100 MHz) is roughly a million times lower than the frequencies of the optical 'carrier' waves ($\sim 10^{14}$ Hz), the difference in optical path lengths from the light source to the two detectors need no longer be restricted to values comparable to an optical wavelength ($\sim 10^{-5}$ cm) as is the case with a Michelson interferometer; light reaching one detector can be retarded with respect to the other by thousands of wavelengths without affecting the correlation (provided the delay is small compared with $c/(100\,\text{MHz}) = 30$ cm).

Technically, the physical quantity actually measured by HBT was not the time-averaged product of the light intensities, but the correlation of the *fluctuations* in intensity at the two detectors. If one represents the instantaneous light intensity received at detector 1 by $I_1(t)$ and the average intensity by $\langle I_1 \rangle$ (which is independent of time for a stable – or so-called stationary – light source), then the fluctuation in light intensity at time t is taken to be $\Delta I_1(t) = I_1(t) - \langle I_1 \rangle$; likewise, the corresponding instantaneous intensity fluctuation at detector 2 is $\Delta I_2(t) = I_2(t) - \langle I_2 \rangle$. HBT measured the time-averaged product of the fluctuations

$$c(d) = \langle \Delta I_1(t) \Delta I_2(t) \rangle = \langle I_1(t) I_2(t) \rangle - \langle I_1 \rangle \langle I_2 \rangle. \qquad (1.12c)$$

This relation differs from that which led to equation (1.12a) only by the last term containing the product of the (time-independent) mean intensities.

The fluctuation in the electric current issuing from a photodetector derives principally from two different origins. The major component is the classical *shot noise* associated with the 'graininess' of electricity, i.e. the fact that charge is transported by discrete units (electrons) rather than by a continuous flow of electrical fluid. The shot noise of one detector is totally independent of the shot noise of the other detector and consequently does not contribute to the correlation function $c(d)$ when the detector outputs are multiplied and time-averaged. The smaller noise component, termed *wave noise*, is associated with the incoming light and arises from the myriad random emissions of electromagnetic radiation by the atoms of the hot source. From atom to atom these emissions vary in amplitude and relative phase at each frequency. Moreover, since the atoms are not all moving with the same velocity relative to the observer, the frequency content of the emissions can also vary from atom to atom as a result of the Doppler effect. The light generated by such a chaotic source (and indeed that is the technical term for it: chaotic light) is a superposition of waves of different frequency with amplitudes and relative phases that fluctuate randomly in time thereby producing the current fluctuations or wave noise at the output of a photodetector. The wave noise at two detectors illuminated by the same source *is* correlated and contributes to $c(d)$.

The above heuristic picture of the functioning of the intensity interferometer is rooted in classical physics at least as far as the light is concerned (although the detectors function by means of the photoelectric effect which is a quantum mechanical process). It did not take long, however, for physicists to wonder about the quantum implications of intensity interferometry and to pose to Brown and Twiss some thorny questions. As a personal observation, I know of many instances where a phenomenon, puzzling from the standpoint of classical physics, received a satisfactory treatment within the framework of quantum physics. In the present case, ironically, a phenomenon happily understandable by means of basic physical optics became a troublesome enigma when examined from the viewpoint of the quantum theory of light.

Although the theory of the intensity interferometer is in principle valid for electromagnetic radiation of any wavelength, there are important practical distinctions in the treatment of radio waves and visible light. As summarised by Hanbury Brown[23]:

Radio engineers, before the advent of masers, thought of radio waves as waves and not as a shower of photons. [Because] the energy of the radio photon is so small and there are so many photons, the energy comes smoothly and not in bursts ... We say that the fluctuations in [the photodetector] output are principally due to 'wave noise' and not to 'photon noise'. By contrast, at optical wavelengths, the energy of the individual photon is much greater and there are relatively few photons, so that we can no longer neglect the fact that the energy comes in bursts. [The] fluctuations ... are due principally to 'photon noise' and not 'wave noise'.

From the standpoint of quantum optics, i.e. the theory of photons, the correlation of wave noise has a surprising, indeed startling, implication: the photons received at the two detectors are correlated. One might think – and many *did* think – that the random emission of classical waves translates into a quantum imagery of randomly emitted photons. And, as logic would seemingly dictate, randomly emitted photons must arrive randomly at separated detectors. But this was *not* the case.

Imagine an experiment in which linearly polarised photons emitted

[23] R. Hanbury Brown, *The Intensity Interferometer* (Taylor and Francis, London, 1976) 4–5.

from the same source and arriving at two separated detectors are counted, and the experimenters keep track somehow of the number of coincident arrivals, i.e. the number of times two photons arrive simultaneously, one photon at each detector, within some specified short time interval. Of course, one expects a certain number of such coincidences to occur accidentally even for completely random arrivals; this can be calculated on the basis of classical statistics. HBT actually did this experiment[24]; they found that the coincident count rate significantly exceeded this background level.

To appreciate just how significant was the observed departure from randomness, it is helpful to consider a superficially different, but conceptually equivalent, experiment and its interpretation. Suppose that instead of measuring the number of coincident linearly polarised photons at two detectors, one measures the time interval (or delay) between consecutive arrivals of photons at *one* detector. For example, one photon arrives and starts a clock; a second photon arrives and stops the clock. The time interval is 8 (in some system of units). The experiment is repeated, and the next time interval is 6. After a sufficiently large number of such cycles have been carried out, the experimenter plots the number of recorded events (the consecutive arrivals) corresponding to a particular delay as a function of delay. Perhaps there might have been 1000 pairs of photons with a time delay of 5 units, 950 pairs of photons with a time delay of 10 units, etc.

The result of such an experiment (Figure 1.10) measures what is termed the conditional probability of receiving a second photon given that a first one was already detected; I will call this simply the conditional probability. The outcome, seemingly difficult to reconcile with one's intuitive expectations, appears astonishing. The probability that two photons of the same polarisation arrive at the detector *simultaneously* (zero delay) is *twice* the probability that they arrive purely randomly. The random arrivals occur for delays significantly greater than a certain time interval corresponding to the temporal (or longitudinal) coherence time t_c of the light source. As the delay time increases from zero to infinity, the conditional probability falls smoothly from 2 to 1 with t_c as the approximate demarcation between correlation and randomness.

To judge from the written recollection of Hanbury Brown the correla-

[24] R. Hanbury Brown and R. Q. Twiss, Correlation Between Photons in Two Coherent Beams of Light, *Nature* **177** (1956) 27.

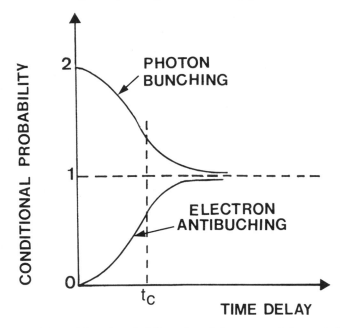

Figure 1.10. The conditional probability of receiving a second thermal photon of specified polarisation after detection of a first is higher for a time interval short with respect to t_c (longitudinal coherence time) than for a comparatively long time interval. In fact, the probability of detecting two photons simultaneously (zero time delay) is twice as high. This behaviour is illustrative of 'photon bunching'. Electrons should display 'antibunching'; the predicted probability of simultaneous detection of two spin-polarised electrons is zero.

tion of photons wreaked havoc on his tranquillity (not to mention his prospects for external funding). In the words of Brown[25]

... if one must think of light in terms of photons then ... one must accept that the times of arrival of these photons at the two separated detectors are correlated – they tend to arrive in pairs. Now to a surprising number of people, this idea seemed not only heretical but patently absurd and they told us so in person, by letter, in publications, and by actually doing experiments which claimed to show that we were wrong. At the most basic level they asked *how, if photons are emitted at random in a thermal source, can they appear in pairs at two detectors*? At a more sophisticated level, the enraged physicist would brandish some sacred text ... and point out that ... our analysis was invalidated by the uncertainty relation ... (Italics added.)

The disturbing question posed to Brown lies at the heart of yet another quantum mystery. I will return to this question later, for it

[25] R. H. Brown, *The Intensity Interferometer, op. cit.*, p. 7.

reverberates like an eerie harmony through the phenomena to be discussed shortly. In the parlance of contemporary quantum optics the pair phenomenon observed by HBT (as well as by others) is termed *photon bunching*. It should be stressed, however, that the graphic imagery of photons grouping together as they propagate through space is misleading. Quantum mechanics does not in general permit us to know the path taken by a particle through space, for any intervention by the observer to 'see' the particle will disturb its motion. The path of a photon is especially problematical, for unlike electrons, photons disappear whenever stopped.

The bunching of light (or – depending on how one wants to regard the phenomenon – excess wave noise) is today an established and noncontroversial fact. The quantum theory of light accommodates with no difficulty the predictions HBT first made on the basis of classical reasoning. Moreover, the experiments that purportedly proved HBT wrong were eventually recognised as being insufficiently sensitive to detect the light correlations. But what, one might wonder, ought to occur in a HBT-type experiment with *electrons*?

This question occupied my attention during much of my time at the Hitachi Research Laboratory. For one thing, more than three decades after the pioneering studies of Brown and Twiss, no comparable experiments with electrons, as far as I was aware, had been attempted, let alone successfully performed. Indeed there seemed to be relatively little discussion of the matter at all. Was it possible to observe HBT-type electron correlations with the beam of an electron microscope? What phenomena would result?

Although analogies with light drawn from the classical domain of physical optics can often provide insights into the quantum interference of electrons, the classical explanation of the optical intensity interferometer provides no help whatever in understanding the corresponding electron interferometer. Electrons, as a consequence of their intrinsic spin $\frac{1}{2}\hbar$ are fermions; no more than one electron can occupy a specified quantum state. Electrons, therefore, cannot form classical waves such as light waves, the quantum description of which entails large numbers of photons with identical quantal properties of energy, momentum and helicity (related to the classical attributes of frequency, wave vector and polarisation). To predict the outcome of an electron HBT-type correlation experiment requires an intrinsically quantum mechanical analysis.

Let us start with a simplified electron correlation *Gedankenexperiment* analogous to the actual photon correlation experiment of HBT.

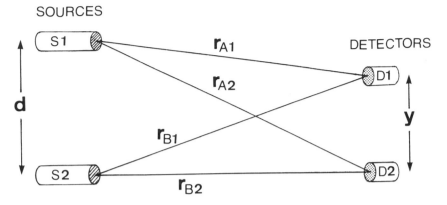

Figure 1.11. Schematic experimental configuration to illustrate electron interference arising from particle indistinguishability. Sources S1 and S2 each emit a spin-polarised electron; detectors D1 and D2 each receive an electron. Since it is not possible to know from which source a detected electron has issued, the amplitudes for the two possible events interfere. The antisymmetry of the electron wave function under particle exchange leads to a joint detection probability $P(D1,D2)$ that vanishes with vanishing source separation d or detector separation y.

Consider two spatially separated compact sources, S1 and S2, that randomly emit electrons all of which have the same energy and spin component, but whose momentum may vary within a narrow range about the forward direction. Sufficiently far from the sources, so that the electrons can be characterised by plane waves, are two detectors, D1 and D2 (Figure 1.11). What is the joint probability as a function of detector separation that D1 and D2 will each simultaneously receive one particle?

From the standpoint of classical physics, where the particles may be thought of as distinguishable, one can reason that the desired probability is the sum of two contributions: (a) the probability that an electron from source S1 goes to detector D1, and an electron from S2 goes to D2, and (b) the probability of the alternative arrangement whereby an electron from S1 goes to D2, and an electron from S2 goes to D1. (We discount the simultaneous production of two electrons from one source and none from the other.) The two-electron wave functions describing the two configurations are

$$\psi_a(1,2) = \psi_{S1}(D1)\psi_{S2}(D2), \tag{1.13a}$$

$$\psi_b(1,2) = \psi_{S1}(D2)\psi_{S2}(D1). \tag{1.13b}$$

For the present discussion, the significant part of each single-electron wave function, represented as a plane wave, is simply a phase factor of the form

$$\psi_S(D) \sim \exp(ikr_{SD}), \tag{1.13c}$$

where r_{SD} is the path length between one of the sources S and one of the detectors D, and $k = 2\pi/\lambda$ is the wave number (or the magnitude of the linear momentum in units of $\hbar$) of an electron with wavelength λ. Calculation of the joint probability

$$P(D1,D2) = |\psi_a(1,2)|^2 + |\psi_b(1,2)|^2 = \text{constant} \tag{1.13d}$$

gives a number independent of the relative separation of the detectors. In other words, the randomly emitted particles arrive totally uncorrelated at the two detectors. The result may seem logically satisfying; the electrons begin life independently at two sources and end up randomly at the detectors. Nature, however, does not always favour logical simplicity based on classical reasoning. The conclusion and mode of thinking are again *wrong*.

I have discussed in the previous section the 'ghostly' correlations inherent in the quantum description of the two-electron Aharonov–Bohm effect with widely separated solenoids. These correlations are inexplicable on the basis of classical physics – but at least they evolved from a special initial condition that could very well be understood in classical terms. The electrons in that example, having been produced in a state of zero total linear momentum, were ever afterward (in the absence of external forces) constrained by the law of momentum conservation to propagate with equal and opposite momenta[26]. This law and the corresponding correlation of momenta apply with equal validity to electrons or to pieces of brick. In a manner of speaking, therefore, a correlation was 'built in' at the outset by a particular initial condition characterisable in both quantum and classical terms. In the present case, however, electrons emerge randomly from what to all appearances are separate sources. What kind of correlation could one possibly expect?

In the quantum mechanical scheme of things the electrons are indistinguishable particles, and it is not possible, even in principle, to

[26] It should be pointed out that the results of the two-electron two-solenoid AB experiments of Section 1.3 would not have been essentially altered if the electron momenta were correlated in some other way than by back-to-back emission. If, for example, the second electron propagated towards mirror M2 (rather than M2′) when the first propagated towards M1, various phase factors would be different, but the nature of the correlation expressed in $P(D1,D2)$ would be the same.

designate from which source a particular electron comes. True, one can always place an electron detector close enough to a source to determine whether it has emitted an electron, but this, of course, is an intervention that alters the motion of the particle (even if the detector were somehow transparent to electrons) and therefore the conditions of the originally intended experiment. The arrival of 'labelled' electrons at specified detectors represents outcomes that can never actually be distinguished. As is well known in such instances (for example, in the case of two-slit interference), one must add the amplitudes, and not the probabilities, for each indistinguishable quantum pathway. But in what way are the amplitudes to be added?

It is precisely at this point that the fermionic nature of the electron – an attribute distinguishing it from the photon in as fundamental a way as electrical charge – enters the analysis. The amplitudes for a given process; for example

$$\text{particle } 1 \rightarrow \text{detector } 1$$

Process a

$$\text{particle } 2 \rightarrow \text{detector } 2$$

and the exchange process

$$\text{particle } 1 \rightarrow \text{detector } 2$$

Process b

$$\text{particle } 2 \rightarrow \text{detector } 1$$

must be superposed with *opposite* signs. This is one example of what is usually termed the spin-statistics connection. The exchange of any two identical particles whose spin is an odd half-integer multiple of $\hbar$ follows the above rule; the aggregate behaviour of such particles is governed by what is known as Fermi–Dirac statistics. (Particles with an even integer spin are classified as bosons, for they are governed by Bose–Einstein statistics; under particle exchange the boson wave function incurs no sign change.)

Why nature works in this way seems to lie outside the framework of quantum mechanics proper; a satisfactory explanation can be made only in terms of the relativistic invariance and microscopic causality of quantum fields. Once, when asked why spin-$\frac{1}{2}$ particles obey Fermi–Dirac statistics, Feynman planned to prepare a freshman lecture on it – but failed. 'You know, I couldn't do it', he said; 'I couldn't reduce it to the freshman level. That means we really don't understand it'[27].

---

[27] D. L. Goodstein, Richard Feynman, Teacher, *Physics Today* **42** (February 1989) 75.

Perhaps this overstates the case somewhat, but the principle is nonetheless a deep one.

In the *Gedankenexperiment* under consideration, the two-electron wave function representing the above two indistinguishable processes is therefore

$$\psi(1,2) \sim \psi_a(1,2) - \psi_b(1,2), \tag{1.14a}$$

where the component wave functions are given in relations (1.13a) and (1.13b). Upon substitution of the single-electron amplitudes (1.13c), the appropriately normalised joint detection probability takes the form

$$P(D1,D2) = |\psi(1,2)|^2 \sim (1/2)[1 - \cos(2\pi y d/\lambda r)], \tag{1.14b}$$

where y is the detector separation, d is the source separation and r is the mean distance of the detectors from the sources (approximately the same for either source to either detector).

The correlation to which the above expression gives rise is very different from that for thermal photons. The joint probability that two electrons arrive at the same location (zero detector separation) is *zero*. This quantum expression of particle avoidance has been termed *antibunching*. The phenomenon of antibunching, which arises from an antisymmetric linear superposition of wave functions as in relation (1.14a), is an example of quantum interference arising, not from space-time differences in alternative geometrical pathways, but from the spin-statistics connection, i.e. from the fact that electrons are fermions.

The joint probability expressed in relation (1.14b) is seen to oscillate repeatedly giving rise to an infinite number of detector locations at which the electron anticorrelation is perfect $(P(1,2) = 0)$. This is a consequence of representing the individual electron wave functions by infinitely extended plane waves instead of by a more realistic wave packet description. An electron source, such as is found in an electron microscope, produces a beam of electrons whose quantum description would include a distribution of particle numbers, energies, linear momenta and spin components. Such a source, like a thermal light source, might also be termed chaotic. Despite the added complexity, the essential feature of antibunching (although not necessarily the oscillations) is predicted to persist in the aggregate electron behaviour.

One dramatic illustration of the anticorrelation of electrons is the conditional probability of electron arrival at a single detector (Figure 1.10). In contrast to the case of thermal light, the probability of detecting a second electron a short time (compared with the beam coherence time t_c) after receipt of a first is suppressed below that expected for totally random particles. For zero time delay, the probability of two electrons arriving together is predicted to be strictly zero – a quantum consequence of the 'minus sign' in electron exchange.

Yet spin and statistics and minus signs aside, how can one understand in some more tangible way the origin of antibunching? For the case of thermal light the correlations amongst photons at least have a classical explanation in terms of fluctuating light waves. But now, in the absence of a classical explanation, the question to HBT comes back even more forcefully in its fermionic version to haunt us: if spin-polarised electrons are emitted at random in a thermal source, how can they *avoid* arriving in pairs at a detector?

Particle indistinguishability and the uncertainty principle help provide a heuristic answer, but one must be careful not to be trapped by the paradox-laden terminology of classical physics. As posed, the question is not physically meaningful, for its premise cannot be substantiated. *Are* the particles emitted randomly? How would one demonstrate this – other than by inferences based on particle detection? Is there any way to determine the instant of emission of each particle without affecting its subsequent motion? Indeed, is there an 'instant' of emission?

As pointed out before, the particles of a beam with an energy uncertainty ΔE do not emerge from their source like mathematical points, but can be represented by wave packets created over a characteristic time interval $t_c = \hbar/\Delta E$, the coherence time. Thus, one could know nothing about the emission time of a hypothetical electron whose wave function is a monochromatic plane wave. If the energy of the beam is not perfectly sharp, but nevertheless defined well enough so that the particles may be assumed to move with speeds close to the mean speed v, an emerging electron will likely be found within a coherence length $l_t = vt_c$.

Two particles whose emission events are separated by a time interval long in comparison to the coherence time are characterised by wave packets that effectively do not overlap; there would then be no quantum interference effects engendered by particle exchange and the spin-statistics connection. These particles arrive, therefore, uncorrelated at the two detectors, and one might think of their emissions as random. However, the wave packets of two particles whose emission events occur in a time interval short with respect to the coherence time can overlap, and the subsequent particle motion can manifest, even in the absence of interactions attributable to forces, the 'ghostly' correlations of particle exchange.

The above response to the question demanded of Hanbury Brown and Twiss must nevertheless be accompanied by cautionary words, for, like any visualisable explanation, it is couched in the language and imagery of classical physics. Such a description can be grossly misleading. To think of the coherence time as providing a definitive criterion for the overlap or nonoverlap of particle wave packets, and hence for the occurrence or nonoccurrence of particle correlation, is not correct. The particles that pass through the intensity interferometer are all part of a single multi-particle system; the correlations, so to speak, are *always* there.

For example, the coherence time of a 150-kilovolt field-emission electron source, a potentially suitable candidate for an electron HBT experiment, is extremely short, about 10^{-14} second. Suppose one were to attempt to observe electron antibunching by measuring, as described before, the conditional probability that a second electron arrives at a given detector at various delay times after receipt of a first electron. It is a near certainty that any experimental attempt with current technology would have to be made with delay times much longer than the coherence time, perhaps two to five orders of magnitude longer. Yet quantum theory shows that, even for delay times orders of magnitude longer than the coherence time, the sought-for correlations will not have vanished entirely – they will still be there, albeit weakly, to be disentangled (by statistical analysis of a sufficiently large number of counts) from the random background events.

The subtleties of correlated electron states and the potential pitfalls of adopting too literally the imagery of wave packets show up strikingly in an experimental configuration combining elements of both the Aharonov–Bohm and Hanbury Brown–Twiss experiments. Consider again the diffraction of electrons, produced by a single source, through two narrow apertures between which is placed an AB solenoid with confined magnetic flux (Figure 1.12). There are now, however, *two* detectors whose outputs are correlated so that the joint probability of detection – in essence, the coincident count rate – can be determined. Since the count rate at one detector has been previously shown (in Section 1.2) to vary harmonically with the magnetic flux, one might well expect that the joint count rate at the two detectors must likewise exhibit a flux-dependent quantum interference effect. Yet surprisingly, if the electrons are correlated, this expectation is not borne out.

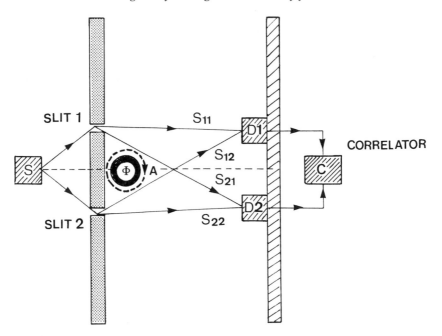

Figure 1.12. Schematic diagram of a hybrid Hanbury Brown–Twiss and Aharonov–Bohm experiment. Electrons emitted from source S pass through slits 1 and 2, around the solenoid (with magnetic flux Φ directed into the page and vector potential field A circulating clockwise) and are received at detectors D1 and D2. The correlated output of the detectors shows an interference effect that depends on both the confined magnetic flux and the fermionic nature of the electron.

Suppose the electron source is again spin-polarised and very nearly monochromatic; the electrons, characterised by plane waves of mean wave number k, have a spread in linear momentum about the forward direction. Then, in view of what has been said before concerning the antisymmetrisation of fermion wave functions, the total amplitude for an electron to pass through each slit and arrive simultaneously at each detector takes the form

$$\psi(D1,D2) \sim \exp(iks_{11})\exp(iks_{22})\exp[i\{\alpha(s_{11})+\alpha(s_{22})\}]$$
$$+\exp(iks_{12})\exp(iks_{21})\exp[i\{\alpha(s_{12})+\alpha(s_{21})\}]. \quad (1.15a)$$

Here s_{ij} is the distance from ith slit ($i = 1,2$) to the jth detector ($j = 1,2$). Thus, the first product of amplitudes represents the process for which electrons from slits 1 and 2 respectively arrive at detectors D1 and D2; the second product of amplitudes represents the exchange process. The phase shifts $a(s_{ij})$, incurred by propagation through the vector potential field of the solenoid, are related to the confined magnetic flux Φ by Stokes' law as follows

$$a(s_{11}) - a(s_{21}) = a(s_{12}) - a(s_{22}) = (e/\hbar c)\Phi \equiv \alpha. \qquad (1.15b)$$

It is then a straightforward matter to show that the (normalised) joint probability of electron arrival at D1 and D2, deducible from relations (1.15a,b), is

$$P(\text{D1,D2}) = |\psi(\text{D1,D2})|^2$$
$$= \tfrac{1}{2}[1 - \cos\{k(s_{11} - s_{21} + s_{22} - s_{12})\}]. \qquad (1.15c)$$

Although quantum interference occurs, there is no trace of the magnetic flux! The magnetic phase shifts for the direct and exchange processes have cancelled.

It appears, at least at first glance, that the two-slit AB effect with electrons correlated by the spin-statistics connection manifests a curious phenomenological reversal *vis-à-vis* the AB effect with momentum-correlated electrons discussed in the previous section. In the latter case, the magnetic flux dependence occurs only in the joint detection probability $P(\text{D1,D2})$; the probability of electron arrival at a single detector, e.g. $P(\text{D1})$, manifests no AB effect. By contrast, the experimental configuration of Figure 1.12 gives rise to a joint probability $P(\text{D1,D2})$ unaffected by the confined magnetic field, although the probability of electron arrival at each detector individually has been shown earlier to vary harmonically with magnetic flux in the following way

$$P(\text{D1}) = \tfrac{1}{2}[1 + \cos\{k(s_{11} - s_{21}) + \alpha\}], \qquad (1.16a)$$

$$P(\text{D2}) = \tfrac{1}{2}[1 + \cos\{k(s_{12} - s_{22}) + \alpha\}]. \qquad (1.16b)$$

If this is the case, then it leads to an extraordinarily puzzling consequence. By arranging experimental conditions so that the geometrical phases are

$$k(s_{11} - s_{21}) = k(s_{22} - s_{12}) = -\pi/2$$

(e.g. symmetrical disposition of D1 and D2 above and below the forward direction) and adjusting the magnetic flux so that $\alpha = \pi/2$, one deduces that $P(\text{D1,D2}) = 1$, $P(\text{D1}) = 1$, $P(\text{D2}) = 0$. How can it be that there is a 100% coincidence count rate, if the individual count rate at one of the detectors is zero!

The origin of the paradox lies in the inconsistent treatment of cor-

tions) by use of faster detectors. Other variables being fixed, the signal-to-noise ratio generally increases as the square root of the total length of time the particle count is maintained. During the 1950s, HBT succeeded in observing the correlations of light beams with a degeneracy on the order of 10^{-3} by collecting data for some ten hours. Although the degeneracy of present electron sources is lower, the use of detectors with response times closer to the electron coherence time t_c would reduce the total counting time needed to achieve a desired signal-to-noise level.

Another promising possibility is to use charged particles other than electrons. For example, with a gas source that produces an intense, collimated, nearly mono-energetic beam of helium ions, one could, in principle, study the correlations of either fermions (such as the $^3\text{He}^+$ ion) or bosons (like the $^4\text{He}^+$ ion) simply by changing the input gas.

Difficult though they may be, these new types of quantum interference experiments will some day be performed, for there are, I believe, few limitations to human ingenuity beyond the laws of physics themselves. And to these rigorous physical laws, I would add another one of an historical nature – namely, the development of new experimental methods nearly always leads to significant discoveries.

* * * * * * * * * * * *

Physicists have had some seventy years to adjust to the discovery of the wave-like behaviour of matter. Time and familiarity often have a way of dulling astonishment, but neither makes the strange processes of the quantum world more visually accessible today than they were before. What is one to make of a description of nature that forbids detailed knowledge of motion, manifests force-free interactions between matter and fields, and gives rise to ghostly correlations between arbitrarily separated noninteracting particles? Schrödinger, the person perhaps most responsible for the wave mechanics of matter, wrote in utter frustration[29]

... that a space-time description is impossible, I reject *a limine*. Physics does not consist only of atomic research, science does not consist only of physics, and life does not consist only of science. The aim of atomic research is to fit our empirical knowledge concerning it into our other thinking. All of this other thinking, so far as it concerns the outer world, is active in space and time. If it cannot be

---

[29] From a letter of Schrödinger to Wilhelm Wien, cited in W. Moore, *Schrödinger* (Cambridge University Press, Cambridge, New York, 1989) 226.

fitted into space and time, then it fails in its whole aim and one does not know what purpose it really serves.

Like all humans, scientists have a deep-rooted need for descriptive explanations; mathematical formalism alone, even if seemingly correct, is somehow insufficient. But quantum mechanics furnishes predictions, not explanations. Perhaps there will come a time when the mysterious wave-like processes inherent in the structure of quantum theory will be unravelled in a causally explicit way – although I rather doubt it. But neither do I find that doubt disturbing. If not purpose, then surely there is at least great satisfaction in a theory of such broad predictive power that opens up for exploration a world beyond the senses where even the imagination can scarcely follow.

2

Quantum beats and giant atoms

2.1 The light from atomic 'pulsars'

When I was at Harvard University many years ago investigating the structure and interactions of the hydrogen atom, I first learned of a remarkable optical phenomenon. I might well have observed it in my own experiments, had it occurred to me to look. But it did not, and I missed seeing at that time a most striking demonstration of the principles of quantum mechanics. By the time I realised this, the apparatus had been 'improved', and the effect would not have been produced under the changed conditions. Nevertheless, like a haunting melody, this curious phenomenon known as 'quantum beats' has often returned to my thoughts to form a significant part of my scientific interests. Had I looked carefully enough at my hydrogen atoms back in the late 1960s, I would have seen them periodically 'winking' at me like the rotating beacon of a lighthouse – like a little atomic pulsar.

Since 1913, when Niels Bohr revealed his semiclassical planetary model of the atom, atomic hydrogen has been a touchstone against which the success of any theory of atomic structure is measured. With but a single bound electron it is the simplest naturally occurring atom of the periodic table – and even today the only atom for which analytically exact theoretical treatments can be provided. (If exotic combinations of particles are included, then positronium, a bound electron and anti-electron (positron) may be considered the simplest atom, for it is a purely electrodynamic system (i.e. not subject to the nuclear strong interactions) containing two apparently structureless particles.) The main legacy of the Bohr theory, retained and refined in the complete quantum mechanics which subsequently followed more than ten years later, is the idea of discrete characteristic states with quantised energies.

That is, a bound electron cannot absorb energy in arbitrary amounts but, in marked contrast to classical theories of the atom, only in quantities that take it from one energy eigenstate to another (from the German 'eigen' = own or particular). Once the discreteness of atomic states and the quantisation of atomic energy were accepted, it followed as a seemingly irrefutable proposition that a free atom, unperturbed by external fields, had at all times to be in one of its eigenstates.

It might be worth noting that there is nothing intrinsically quantum mechanical about the concept of 'quantisation' – the very feature that gave the 'new' mechanics its name. Indeed, the attribute of discreteness of allowed values is encountered in other systems of a purely classical nature as, for example, the quantised oscillation frequencies of a vibrating string, membrane or air column. Quantisation is a consequence of the imposition of boundary requirements – and this can occur in classical or quantum mechanics. Frequency and energy are closely linked in quantum mechanics by Einstein's relation, $E = h\nu$, in a way without parallel in classical physics. Nevertheless, if I had to give a one-word synopsis of what is phenomenologically unique in quantum mechanics, it would be 'interference', not 'quantisation'; but this is a matter of personal opinion. The phenomenon of quantum beats is the embodiment of interference – the ultimate two- (or more) slit particle interference experiment packed into the diminutive volume of a single atom.

One objective of my Harvard experiments was to probe the structure of the hydrogen atom more thoroughly than had been done before. Motivating every 'hydrogen watcher' is the hope of finding a discrepancy with theory, for, although there is satisfaction in confirming quantum mechanics, it would be far more exciting to disprove it. Bohr's prediction of the spectrum of electronic energy levels

$$E_n = -Ry/n^2 \quad (n = 1, 2, 3, \ldots) \tag{2.1a}$$

for a particle bound in a Coulomb potential $[V(r) = -e^2/r]$ was correct as far as it went. Here, the Rydberg constant, Ry, defined by

$$Ry = 2\pi^2 m e^4/h^2 \sim 13.6 \, \text{eV}, \tag{2.1b}$$

where h is Planck's constant, and e and m are the electron charge and mass, sets the scale of atomic energies. However, the energy level structure of a real hydrogen atom is more complex.

For one thing, the intrinsic spin $\frac{1}{2}\hbar$ (recall $\hbar = h/2\pi$) of the electron gives rise to an electron magnetic dipole moment. The magnetic moment of a classical particle is proportional to the angular momentum of the particle and inversely proportional to its mass. This is also the case for the quantum mechanical electron although the proportionality constant is a factor 2 larger than that deduced from classical mechanics. The electron magnetic moment emerges in a natural way from the Dirac relativistic equation of motion.

From the perspective of an observer at rest in the laboratory, the proton in a stationary hydrogen atom is practically at rest. Actually, the electrostatically bound electron and proton orbit in binary-star fashion about their common centre of mass. The location of the centre of mass is almost coincident with the location of the proton, the more massive particle of the pair by a factor of nearly 1840. As seen from the electron rest frame, however, the proton is in orbit about the electron (just as an Earth-bound observer sees the diurnal passage of the Sun). This picture, of course, is drawn from the imagery of classical physics which, when pushed too far, can give misleading, if not totally erroneous, results; quantum mechanics does not ordinarily allow us to imagine electron or proton trajectories within an atom. Still the picture can be useful at times.

The orbiting proton (in the electron reference frame) constitutes an electrical current which produces at the electron site a magnetic field proportional to the electron orbital angular momentum (in the laboratory reference frame). Depending upon whether the electron spin and orbital angular momenta are parallel or antiparallel to one another (the only allowed possibilities for a spin-$\frac{1}{2}$ particle), the interaction between the electron magnetic moment and the local magnetic field can slightly augment or diminish the electrostatic (Coulomb) energy of an atomic state.

Each Bohr energy level of given principal quantum number n actually comprises $2n^2$ degenerate states – i.e. states of the same energy – distinguished by quantum numbers designating their orbital angular momentum (L), component of orbital angular momentum along an arbitrarily chosen quantisation axis (M_L), and component of spin along that same axis (M_S). Thus this fine structure or spin–orbit interaction splits the Coulomb energy of states with nonzero angular momentum quantum number L into two close-lying levels. The exact amount of splitting depends on the angular momenta of the states involved, but to a good approximation it is smaller than the electrostatic energy by the square of the so-called Sommerfeld fine structure constant, $\alpha_{fs} = e^2/\hbar c \sim 1/137$, and the first power of the principal quantum number, i.e.

$$\Delta E_{\text{fs};n}/E_n \sim \alpha_{\text{fs}}^2/n \sim 5 \times 10^{-5}/n. \tag{2.1c}$$

The hydrogen fine structure intervals divided by Planck's constant correspond to Bohr frequencies that generally fall in the microwave or radiofrequency range of the electromagnetic spectrum.

The proton, like the electron, is also a spin-$\frac{1}{2}$ particle with a magnetic moment. However, the relation between the proton magnetic moment and spin is not as simple as for an electron. The reason for this is that the electron, as far as one presently knows, is a true 'elementary' particle with no internal structure, in contrast to the proton which is thought to be a composite of three more elementary particles known as quarks and has a complex internal structure. The hyperfine, or spin–spin, interaction between the proton and electron magnetic dipole moments splits each fine structure level further. This hyperfine splitting again depends on the quantum numbers of the states in question, but to good approximation is smaller than the fine structure splitting by the ratio of the electron and proton masses; thus

$$\Delta E_{\text{hf};n}/E_n \sim (m/m_{\text{p}})\alpha_{\text{fs}}^2/n. \tag{2.1d}$$

Besides the Coulombic, spin–orbit, and spin–spin interactions, which have analogues in classical electromagnetism, there are processes that have no direct counterparts in classical physics. These involve the interaction of the bound electron with the 'vacuum'. Classically, a vacuum is empty space; not so in quantum physics. The quantum electrodynamical vacuum is a roiling sea of ephemeral (or virtual) particles of matter and light (photons) that can affect the properties of real particles although their own existence is so short-lived as to preclude the possibility of direct observation.

One effect of the vacuum on atoms is quite well known although perhaps not thought about in this context: the spontaneous emission of light. The fluctuating virtual electromagnetic fields of the vacuum stimulate excited atoms to undergo transitions to lower energy states, thereby emitting real photons. Thus, the interaction of an atom with the vacuum results in a finite lifetime of the excited atomic states. There are also other more exotic processes that affect the atomic energies.

A bound electron, for example, can interact with the vacuum to emit and then immediately reabsorb a photon; this process alters what is known as the electron self-energy. The electron self-energy, the calculation of which yields an infinitely large value, is not measurable. However, the *difference* in self-energy values between a free electron and one bound in a hydrogen atom *is* calculable and measurable. Another such process, referred to as vacuum polarisation, involves the emission by the atomic nucleus of an electron–positron pair, the immediate mutual annihilation of this pair to produce a photon, and the absorption of this photon by the bound electron. The net effect of these (and other) virtual processes is to shift the energy levels of different states by different amounts. The most notable shift is between the states designated $n\mathrm{S}_{\frac{1}{2}}$ and $n\mathrm{P}_{\frac{1}{2}}$ which, according to the relativistic quantum theory of Dirac, are predicted to have equal energy in the absence of vacuum processes[1].

The atomic beam experiments in which I was engaged were undertaken to explore a wide range of hydrogen fine structure and hyperfine structure intervals and the shifts induced by vacuum processes. I liked to think of this research as my 'Nobel Prize Project' – *not* because the work was destined to win one, but because each principal experimental ingredient of the project had already earned someone else a Nobel Prize. For the development of the 'molecular ray method' or beam, Otto Stern received the Prize in 1943. I. I. Rabi won it in 1944 for his 'resonance method', the discovery that one can induce transitions between quantum states of a nucleus by irradiation with a magnetic field oscillating at the appropriate Bohr transition frequency. The procedure, as I employed it, worked just as well with an oscillating electric field applied to electronic fine structure states of the hydrogen atom. Investigation of the hydrogen fine structure earned Willis Lamb the Prize in 1955; the quantum electrodynamic displacement of S and P states bears his name (Lamb shift). Much later (in 1990) Norman Ramsey, a former Rabi student, was to receive the Prize for a modification of the resonance method whereby *two* spatially separated, but coherently oscillating, radiofrequency fields allowed one to measure nuclear energy level intervals with high precision. The use of this technique, whose theoretical possibilities I studied at great length, significantly improved the precision with which the hydrogen Lamb shift could be determined. In addition, if the creators of the theoretical underpinnings of the experiments were also to be acknowledged, then, of course, the list of Nobel Laureates must include Bohr (1922), for 'the investigation of the structure of atoms, and of the radiation emanating from them',

[1] The fine structure states of atoms are labelled by (i) the principal quantum number n specifying the electronic manifold, (ii) a letter indicative of the orbital angular momentum L in units of $\hbar$ (S = 0, P = 1, D = 2, F = 3, G = 4, and so on in alphabetic sequence), and (iii) a numerical subscript giving the total (i.e. orbital+spin) electron angular momentum J in units of $\hbar$.

Heisenberg (1932), 'for the creation of quantum mechanics', and Schrödinger and Dirac (1933), 'for the discovery of new productive forms of atomic theory'. Newton once remarked that, if he saw farther than most, it was because he stood on the shoulders of giants (... a comment uttered during an 'unusual fit of modesty' according to one of my historian colleagues). In any event, the predecessors upon whose achievements I relied, had no mean stature, either.

Since all hydrogen states, except for the ground state[2], are unstable and decay radiatively to some lower state(s), one could monitor the effects of external perturbations on them by the corresponding increase or decrease in light emission at the appropriate wavelength. The use of a *fast* atomic beam – a beam in which the atoms move through the apparatus at roughly a hundredth the speed of light – greatly facilitates such spectroscopic measurements. Produced at one location, the atoms rapidly traverse various chambers containing the electromagnetic fields (oscillating at radio or microwave frequencies) for probing desired energy level intervals; the atoms continue past a detecting window through which the fluorescent photons (the spontaneous decay radiation) can be counted, thereby providing a measure of the number of atoms remaining in the states of interest. When the frequency of an oscillating field corresponds to the Bohr transition frequency for a pair of atomic states[3], the probability of a transition into or out of the states is greatest, and the photon count rate is maximally affected. (Whether a transition occurs into or out of atomic states coupled by the oscillating fields depends on the relative populations of these states and the time of exposure to the fields.) Thus, use of a fast beam allows for separate regions of creation, spectroscopy and detection of short-lived states.

One might wonder, however, how a *neutral* hydrogen atom can be brought up to a speed 8000 times that of a passenger jet! After all, unlike the charged particles in high-energy accelerators, a neutral particle cannot be accelerated by electric or magnetic fields. The trick is first to accelerate a beam of *protons* to the desired speed. The protons were produced in an ion source – essentially a cylindrical glass tube (fed

---

[2] The $1S_{1/2}$ ground level actually comprises four states distributed within two hyperfine components designated by the total (electron orbital + electron spin + nuclear spin) quantum number F. The energy interval between the lower, or more tightly bound, F = 0 state (the true ground state) and the three degenerate F = 1 states (with magnetic quantum numbers $M_F = -1, 0, +1$) corresponds to a frequency lying in the microwave region of the spectrum – the 1420 MHz line of great importance in radioastronomy and astrophysics.

[3] The Bohr frequency ν_{12} for a quantum transition between a state with energy E_1 and a state with higher energy E_2 is given by $h\nu_{12} = E_2 - E_1$.

H_2 gas) inserted through the coil of a powerful radiofrequency oscillator which dissociated the molecular hydrogen into a plasma that gloriously radiated Harvard's crimson colour (the Balmer lines of excited H atoms[4]). Extracted from the plasma and then electrostatically accelerated under a potential difference of about 20 000 volts, the protons impinged on a thin carbon foil a few hundred atoms thick capturing electrons as they shot through virtually unaffected in their forward motion. Now the accelerated proton beam had become a beam of fast-moving hydrogen atoms distributed over a broad range of quantum states.

There were many aspects to the experiment that were challenging, but few more frustrating than these foils which had a tendency to 'burn through' just when the collection of data seemed to be going well. Replacing them was a time-consuming affair, for the accelerator had to be shut down and opened to the atmosphere, after which began the tedious task of separating the ultra-thin carbon foils from glass microscope slides on which they were mounted by the manufacturer, and of then remounting them (without crumpling or breaking) on a frame to be suspended in the path of the beam. Useful though they had been, I was not unhappy to dispense with the whole business of carbon foils and use an indestructable gas target that accomplished the same task with less aggravation. I had not realised, unfortunately, that what was potentially the most interesting part of the experiment was literally thrown away!

The characteristic feature of the random decay of independent systems – whether alpha particle decay of atomic nuclei or radiative decay of excited atoms – is the exponential variation in time. This is the inevitable result of a decay process in which the number of particles decaying at any moment is proportional to the number of particles present:

$$dN/dt = -N/T. \qquad (2.2a)$$

Here $1/T$ is the characteristic decay rate of the particular process; T is said to be the particle 'lifetime' (for the given mode of decay), but this is a statistical quantity referring to the whole ensemble of particles, and not an attribute of an individual particle which may live considerably longer or shorter than an interval T. The differential equation (2.2a) is

---

[4] The crimson colour of excited atomic hydrogen arises from the spontaneous emission of red photons in the transition $n = 3 \rightarrow n = 2$, and blue photons in the transition $n = 4 \rightarrow n = 2$.

readily integrated. After a time interval t, the ratio of the number, $N(t)$, of remaining states to the number, N_0, of initial states is

$$N(t)/N_0 = \exp(-t/T). \tag{2.2b}$$

The lifetime T, therefore, is the time interval after which the relative population of decaying particles has dropped to $e^{-1} \sim 0.37$.

As I mentioned before, the belief was widespread that an atom unaffected by external perturbations had to be in one of its allowed energy eigenstates. Thus, the beam of fast H atoms emerging from the carbon foil would contain what one could describe as a *mixture* of states. A complete description of such a mixture would entail a tabulation of the statistical frequencies or probabilities with which each hydrogenic state appears; perhaps something like 80% 1S states, 10% 2S states, 5% 2P states, etc. (if one limits the description to the orbital states of different electronic manifolds). As the beam leaves the foil, the excited states decay in time at rates that depend on the principal and orbital angular momentum quantum numbers. For example, the lifetime of a 2S state is about $\frac{1}{7}$ second (effectively infinite on the time scale of atomic processes[5]), whereas that of a 2P state is 1.6 nanoseconds ($1\,\mathrm{ns} = 1 \times 10^{-9}$ second). The uniform motion of the beam converts the decay over a time interval to decay over a space interval.

Suppose one examined the light output from the decaying 4S states (lifetime ~ 230 nanoseconds) by placing a filter in front of a photodetector to block all radiation except for the blue Balmer β light (4S to 2P transition) of wavelength about 4860 ångströms (486 nanometres). As a function of distance x from the foil, the Balmer β light intensity, proportional to the number of decaying atoms in the 4S state, would be expected to fall off exponentially in accordance with relation (2.2b) as follows

$$I(x)/I(0) \sim \exp(-x/vT), \tag{2.2c}$$

where the left-hand side is the relative light intensity at x, and v is the beam velocity. To be sure, a photodetector surface is not a mathematical point; the observed signal in an actual experiment comprises photons from decays spanning a range of locations along the beam and

---

[5] Ordinarily, a single photon is emitted when the bound electron undergoes a transition to a lower energy state. However, an electron in a 2S state can decay only to a 1S state, a process that involves no change in orbital angular momentum. Since an emitted photon would have to carry away one unit of angular momentum, a one-photon 2S → 1S transition is forbidden by angular momentum conservation. The 2S state can decay by emission of two photons; this process has a low probability of occurrence, and the 2S lifetime is correspondingly long.

striking different points of the surface of the detector. If one is principally interested – as I was at the time – in maximising the light output, it is advantageous to use a photodetector with a wide window, survey a broad segment of the atomic beam, and count photons of all polarisations. The signal is then suitably given by relation (2.2c) with appropriate averages made over beam length and detector surface. Nevertheless, neither I nor my experimental colleagues had any doubt that the light emitted from a narrow segment of the beam followed the exponential decay law.

Except that it didn't[6]! Or rather it did if one observed all polarisations equally, and it did not if the light was detected through a polariser – e.g. a simple sheet of polaroid film. In the latter case the light intensity oscillated with distance from the foil indicating that the atoms, like miniature beacons, were in some way turning on and off coherently. As the mean distance between atoms in the beam was far greater than a characteristic atomic size, and as the production of atoms by proton impact on carbon apparently took place independently and randomly, there was no reason to believe that different atoms in the beam could in any way cooperate with one another. The oscillating light output reflected in a profound way an oscillatory process intrinsic to each atom – but with all the atoms in synchrony. Like the build-up of a pattern of interference fringes by single electrons as described in Chapter 1, the observed intensity oscillations *could* have been produced one atom at a time – provided one had the patience to collect enough photons.

What were these oscillations? Why did they appear only in polarised light? Why did the standard description of radiative decay not work? Ordinarily applicable in all instances of incoherent particle preparation and decay, relations (2.2a) and (2.2b) do not take account of the uncertainty principle.

A proton moving at roughly 10^8 cm/s will pass through a 10^{-6} cm thick fixed carbon foil in a time interval of about 10^{-14} s. At some point in that short time interval a hydrogen atom is created. As I pointed out before, an uncertainty in the time of production Δt implies an uncertainty in the energy of the system: $\Delta E \sim h/\Delta t$. In the present case this energy uncertainty is *larger* than all the hydrogen fine structure and hyperfine structure energy splittings! The Bohr frequency for the largest fine structure splitting (that of 2P states) is about 10^{10} Hz; the largest hyperfine structure splitting (that of the 1S states) is about 1.4×10^9 Hz.

[6] J. Macek, Interference Between Coherent Emissions in the Measurement of Atomic Lifetimes, *Phys. Rev. Lett.* **23** (1969) 1.

However, the energy uncertainty (expressed in frequency units) of the foil-excited atoms is $\Delta \nu = \Delta E / h = 1 / \Delta t \sim 10^{14}$ Hz.

In effect, the experimenter cannot know to what state any atom has been excited. Of course, if he were to intervene in some way to measure the precise energy of each atom in the beam, this energy would turn out to be one of the energy eigenvalues, but then the quantum beats would disappear. If the experimenter does not measure the energy of an atom, then that atom cannot, even in principle, be thought of as being in a well-defined, albeit unknown, quantum state. The appropriate quantum description must entail a linear superposition of all allowable energy states that give rise to the same final condition – i.e. decay to a specified lower state with emission of a photon falling within the 'passband' of the measuring apparatus. (If the passband were sufficiently narrow that detected photons came from one particular state of a linear superposition of excited states, then the energy of the atom would be known, and the quantum interference would vanish.)

Let us consider the example of a four-state atom with two close-lying excited states such as that shown in Figure 2.1a. The excitation is assumed for the time being to be nearly instantaneous. The atom goes from the initial state g (assumed here to be the ground state) to some final state f by quantum pathways $g \rightarrow e_1 \rightarrow f$ or $g \rightarrow e_2 \rightarrow f$. Since, under the circumstances of the experiment, these paths are indistinguishable,

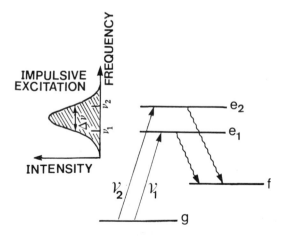

Figure 2.1(a). Energy level diagram of an atom with ground state g, two close-lying excited states e_1 and e_2, and a lower final state f. An impulsive excitation with frequency components corresponding to the Bohr transition frequencies from g to e_1 and e_2 can drive the atom into a linear superposition of its excited states from which the excited electron radiatively decays to lower state f.

one must add the probability amplitude for each. Suppose the probability amplitude for transition from the ground state to an excited state
e_i is a_i ($i = 1,2$) and the corresponding amplitude for a transition from
the excited state e_i to the final state f is b_i ($i = 1,2$). Were it possible to
'turn off' the interaction of the atom with the vacuum, then, during the
time the atom is excited, it would evolve freely in the absence of all
external forces and potentials. From the quantum mechanical equation
of motion (Schrödinger or Dirac equation) one can readily deduce that
the probability amplitude for free evolution in a state of energy E is
proportional to the phase factor $\exp(-iEt/\hbar)$. It is the interaction with
the vacuum, however, that induces the radiative transition to state f that
makes detection of the quantum beat possible. The theoretical effect of
the vacuum can be calculated rigorously by means of quantum electrodynamics; the end result – more or less consistent with our intuition
– is that, besides the free-evolution phase factor of magnitude unity
(no change in number of atoms in the state), there is a decay factor
$\exp(-t/2T)$ representing a loss of atoms from the excited state (with a
characteristic lifetime of T)[7]. Thus, the total probability amplitude for
each pathway is expressible as

$$A(g - e_1 - f) \sim a_1 b_1 \exp(-iE_1 t/\hbar) \exp(-t/2T), \qquad (2.3a)$$

$$A(g - e_2 - f) \sim a_2 b_2 \exp(-iE_2 t/\hbar) \exp(-t/2T). \qquad (2.3b)$$

The probability for the transition of the atom out of g and into f with
emission of one photon at time t is then

$$P(t) = |A(g - e_1 - f) + A(g - e_2 - f)|^2, \qquad (2.3c)$$

which takes the form

$$P(t) \sim [|a_1 b_1|^2 + |a_2 b_2|^2 + 2|a_1 a_2 b_1 b_2| \cos\{(E_2 - E_1)t/\hbar + \phi\}]$$
$$\times \exp(-t/T). \quad (2.3d)$$

The phenomenon of quantum beats is in essence an interference
effect in *time* rather than in space; the temporal sequence of transitions
between two (or more) sets of internal energy states of the atom is
analogous to the spatial pathways through one or the other of two slits in
the free electron 'Young's fringes' experiment described earlier. The
transition probability – and therefore the photon count rate – decays in
time as a *modulated* exponential (Figure 2.1b). The 'beat' frequency in

---

[7] The *amplitude* contains a factor $\exp(-t/2T)$ in order that the *probability* (proportional
to the square of the amplitude) diminish as $\exp(-t/T)$.

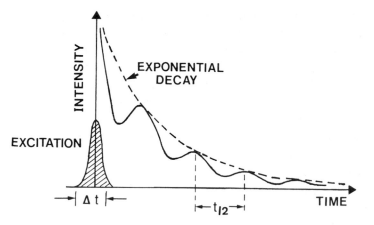

Figure 2.1(b). The spontaneous emission from atoms prepared as shown in Figure 2.1(a) exhibits an oscillatory decay in time with a period t_{12} inversely proportional to the Bohr transition frequency between states e_1 and e_2: $t_{12} = 1/(\nu_2 - \nu_1)$. The spontaneous emission from atoms in an incoherent mixture of the two excited states would decay exponentially in time as shown by the broken line.

the quantum interference term corresponds to the Bohr frequency of the excited states; it may thus be seen that the measurement of quantum beats can afford useful spectroscopic information about the energy level structure of an atom or molecule. About this aspect I will have more to say later. The relative phase ϕ that appears derives from the phases of the excitation and decay amplitudes which can be complex numbers.

The system of percussionally excited atoms in a linear superposition of two quantum states with an oscillating relative phase factor $\exp\{-i(E_2 - E_1)t/\hbar\}$ may be likened to a system of synchronously precessing electric dipoles. Although this picture is a classical one, it helps account for some of the features of the quantum beat experiment. The maximum intensity of the radiation from any one dipole sweeps past the detector at the angular frequency $\omega = (E_2 - E_1)/\hbar$. However, for an ensemble of randomly oriented (although synchronously oscillating) dipoles, the net signal is modulated only when light of a particular polarisation is observed; the precession can alter the distribution of the radiation, but not the total amount generated (which decreases exponentially in time). This is reflected as well in the quantum mechanical expression (2.3d) which characterises a transition induced by a particular component of the electric dipole of the atom (a vectorial quantity), and therefore the emission of a photon of specified polarisation. If one adds together comparable expressions for *all* possible electric dipole transitions from the given excited states to *all* substates (if there are more than one) of the final level f – which is tantamount to observing light of all polarisations – the quantum beat term vanishes, a consequence of the quantum rule known as the Wigner–Eckart theorem.

The picture of precessing dipoles also helps one visualise what should transpire as one increases the time interval over which the excited states are prepared. If the duration of the excitation T_e is short compared to the period of dipole precession $2\pi/\omega$, the dipoles precess together and emit radiation in phase. As T_e lengthens, however, different dipoles are set precessing at increasingly later times and emit radiation increasingly out of phase with that emitted by dipoles established earlier. Once T_e is comparable with, or longer than, $2\pi/\omega$, the precession of the dipoles is no longer synchronised, the radiation is no longer in phase, and the quantum beats are washed out. As a quantitative criterion, the range ωT_e of precession angles spanned by a system of dipoles created over a period T_e must be small compared with unity (1 radian) if the dipoles are to precess synchronously. Thus for the beats to be observable, one would expect the inequality

$$\omega T_e \ll 1 \qquad (2.4a)$$

to hold. This criterion also follows readily from the quantum mechanical analysis. With $\omega = (E_2 - E_1)/\hbar$, and $T_e \sim \hbar/\Delta E$, one recognises in relation (2.4a) a consequence of the uncertainty principle

$$E_2 - E_1 \ll \Delta E. \qquad (2.4b)$$

The uncertainty in the energy of the bound electron must be greater than the energy interval separating the excited states; otherwise, the atom would be in a definite excited state, and beats would not occur.

Excitation of atoms by a carbon foil is not the only means by which quantum beats can be produced. Indeed, in many respects it is advantageous to excite the atoms optically, for example with a pulsed laser. Whereas electron capture from a carbon foil gives rise to many excited states simultaneously, excitation with a tunable laser permits one to select specific excited states of interest. Another nice feature about optical excitation is that the excitation amplitudes (the a_i) can be determined precisely; the theory of the interaction of atoms with light is, if not simple, at least well understood. By contrast, the capture of electrons by proton impact on a solid target of multi-electron atoms is a more difficult process to treat theoretically.

The theory of quantum beats produced by light pulses predicts some unusual optical effects that, as far as I know, have yet to be demonstrated experimentally. They must exist, however, if our understanding of the interaction of atoms with light is correct.

Before the development of powerful pulsed lasers, the light sources used to excite atoms were 'weak' in the sense that the majority of exposed atoms remained in their ground state; that is, the probability of a transition was low and the lifetime of the ground state was long (in principle infinitely long in the absence of radiation). An atom that absorbs a photon from a weak light pulse undergoes effectively one transition to the excited states and subsequently – after passage of the pulse – decays by spontaneous emission to lower states (including the ground state). If the light pulse, like the carbon foil excitation, is short compared to the dipole precession time $(2\pi/\omega)$, then – as one would expect – the probability of spontaneous emission is negligibly small throughout passage of the pulse.

If a light pulse is sufficiently intense, however, it can stimulate atoms to *absorb and emit* photons during its passage. Indeed, many such cycles of excitation by light absorption followed by stimulated emission to the ground state could occur over the duration of one pulse. Thus, as a consequence of strong light excitation, the atomic ground state acquires a finite lifetime T_p inversely proportional to the light intensity at the transition frequency. T_p is the inverse of the excitation rate; it is the 'pumping' time, the mean time between the successive absorption and stimulated emission of a photon. When the pulse has passed, the excited atoms again decay exclusively by spontaneous emission.

What effect should all this cycling back and forth between ground and excited states have on the quantum beats? Very little, one might imagine. After all, the observed beats occur in the *spontaneous* emission from *freely* precessing atomic dipoles *after* passage of the light pulse. This is certainly true if the light pulse is short compared with the precession time. However, a theoretical study of the effect of increasing the pulse duration produced a most surprising result.

One might expect, in view of the reasoning behind relation (2.4a), that the contrast[8] of quantum beats should diminish and ultimately vanish as the duration of the light pulse, T_e, exceeds the precession time characteristic of the excited states. This was indeed the case for light

---

[8] The contrast of the beats is analogous to the visibility of the fringes of an interference pattern. In an expression such as (2.3d) it is the ratio of the coefficient of the time-dependent quantum interference term to the sum of the two time-independent terms representing spontaneous emission from the individual excited states.

pulses of weak to moderate intensity. However, when the intensity of a long ($T_e > 1/\omega$) light pulse was increased sufficiently so that the ground state lifetime was short ($T_p \ll 1/\omega$), the quantum beats reappeared strongly! How was it possible – to refer again to the classical analogy – for apparently randomly phased dipoles to emit light synchronously?

The explanation of this baffling phenomenon turned out to be simple, but subtle. Because of the frequent cycles of excitation and stimulated emission, the precession of the dipoles is interrupted so often that their overall dispersion in phase angle remains small. The system of randomly excited and de-excited atoms resembles somewhat the 'random walk' of a drunkard through the woods: he bumps into trees, falls down, gets up and starts off again – sometimes in the original direction, sometimes in the opposite direction. At the end of a certain time, he has progressed in a random direction from his point of origin by a distance that varies as the square root of the number of steps.

During the passage of the light pulse in a time interval T_e, the number of successive absorption and stimulated emission processes that occur is approximately $N = T_e/T_p$. Each time an atom is re-excited from the ground state, the corresponding dipole can precess either in the original sense or in the opposite sense. Over the time T_p that the atom is excited (before another stimulated emission to the ground state occurs), the corresponding dipole precesses through an angle $\theta \sim \omega T_p$. The dispersion in phase over the whole system of atoms, like the mean displacement in a random walk problem, is $\Delta\theta = N^{\frac{1}{2}}\theta$. The criterion for the appearance of quantum beats, $\Delta\theta \ll 1$ radian, is then expressible as

$$T_e/T_p > (\omega T_e)^2. \tag{2.4c}$$

Relation (2.4c) is equivalent to that of (2.4a) when the excitation is weak and the ground state lifetime long, $T_p > T_e$. However, even when $\omega T_e > 1$, a pulse of long duration should still lead to quantum beats if the pumping time (i.e. ground state lifetime) is made short enough by an intense illumination. This would indeed be an interesting phenomenon to observe.

* * * * * * * * * * * *

By the time my atomic beam experiments were completed, the results did not show any discrepancy with quantum mechanics, and I thought I knew all I ever wanted to know about hydrogen . . . at least for a while. I soon realised, however, that atoms similar in electronic structure to hydrogen have an intrinsic interest all their own, especially when they are so large that *one* such atom could accommodate some 50 000 000 'ordinary-sized' atoms in its volume!

Here was a whole new domain of atomic physics to explore – through a portal opened by pulsed lasers and quantum beats.

2.2 Anomalous reversals

The atomic hypothesis has been around for some two millenniums. Despite compelling evidence provided by the study of chemistry and the kinetic properties of gases, acceptance of the actual existence of atoms was strongly resisted by reputable scientists – including some of the foremost chemists of the time – even as late as the first decade of the twentieth century.

One problem, of course, is that atoms are ordinarily much smaller than the least object that could be seen through a microscope. Physical optics teaches us that one cannot resolve objects of a size inferior to the wavelength of the light used for viewing[9]. The characteristic diameter of an atom is some three orders of magnitude smaller than the wavelength of visible light. As pointed out earlier, the wavelength of electrons in an electron microscope can be a fraction of an atomic diameter; with such a microscope one can (in a manner of speaking) 'see' structures interpretable as an aggregate of atoms. But this is a recent development. No one at the turn of the twentieth century could have conceived of seeing an atom.

The scale of molecular size was already roughly known in the nineteenth century by means of chemical experiments or kinetic experiments to determine Avogadro's number from macroscopic quantities of matter. I can recall, as a student, having to estimate the length of some kind of oleic molecule from the amount of substance required to form a monomolecular film on an aqueous substrate. (How the instructor could be certain that the film was monomolecular was never made clear to me!) By knowing the molecular formula, I could then estimate the size of a carbon atom. No theory based on classical physics, however, was able to predict the characteristic size of an atom or molecule. The reason, in short, is that Planck's constant was not known.

---

[9] This is a consequence of light diffraction, the light usually being observed far (i.e. many wavelengths) from the diffracting object. Less well known, however, is the fact that structures smaller than a wavelength can be resolved when observed very close to the diffracting object (under so-called 'near-field' conditions).

Bohr's semiclasscial theory of the atom in 1913 was the first to provide a natural scale of atomic size, the Bohr radius a_0,

$$a_0 = h^2/4\pi^2 me^2 \sim 5 \times 10^{-9}\,\text{cm} \tag{2.5a}$$

in terms of Planck's constant and the electron charge and mass. The Bohr theory showed that the characteristic size of the orbit of an atomic electron in an energy state of principal quantum number n is

$$r_n = n^2 a_0. \tag{2.5b}$$

Before h entered the physicist's lexicon of physical constants, the only natural length scale that could be constructed from known particle attributes and universal constants was the 'classical electron radius'

$$r_0 = e^2/mc^2 \sim 3 \times 10^{-13}\,\text{cm}, \tag{2.5c}$$

which was orders of magnitude smaller than the size of atoms inferred from experiment; it is more characteristic of the size of the atomic nucleus.

The n^2 dependence of atomic size implies, however, that highly excited atoms are *not* necessarily small – that in fact they are larger than some of the observable and manipulatable objects still adequately treated by the laws of classical physics. Were it possible to raise an electron to the $n = 100$ level, the orbital radius would be $10^4 a_0$, or about 0.5 micron (recall: $1\,\mu\text{m} = 10^{-4}\,\text{cm}$), which is already on the order of the size of some bacteria. Pulsed lasers indeed make such excitations possible; for example, under laboratory conditions barium atoms[10] have been excited to electronic levels on the order of $n = 500$ with a corresponding Bohr radius of $12.5\,\mu\text{m}$. Atoms in comparable states of excitation also occur naturally in the interstellar medium. Note that human red blood cells have diameters of $6–8\,\mu\text{m}$, and most other human cells fall in the range of about $5–20\,\mu\text{m}$.

There is something fascinating about an atom that one should be able to 'see'! Unfortunately, it is not possible to do anything of the kind. To see the atom requires that one illuminate it and that the atom scatter the light to a detector. However, a highly excited atom – generally termed a Rydberg atom – is markedly sensitive to its environment; the least perturbation will likely de-excite or ionise it. The interaction of an atom with an electric field, for example, ordinarily depends on the atomic polarisability which is a measure of the extent to which the field can

---

[10] J. Neukammer *et al.*, Spectroscopy of Rydberg Atoms at $n \sim 500$, *Phys. Rev. Lett.* **59** (1987) 2947.

displace electric charge from its equilibrium distribution. Polarisability has the dimension of volume, and one might expect that the atomic polarisability would scale as the cube of the orbital radius, or as n^6. This is not strictly the case – the scaling goes approximately as n^7 – but it provides a good indication of the difficulty faced by someone wishing to probe, but not destroy, a Rydberg atom[11]. The polarisability of a hydrogen atom in the level $n = 100$ would be over one million million times greater than that of the atom in its ground state.

What makes highly excited atoms particularly interesting to study is, among other reasons, that they are systems at the threshold between the quantum world and the classical world. This is the 'anti-twilight zone', so to speak, where quantum strangeness is expected to merge into classical familiarity by means of the correspondence principle. Since the electrostatic force that binds the electron to the nucleus has the same inverse-square distance dependence as the gravitational force that binds the planets to the Sun, one might think of an atom with the outer valence electron excited into a Rydberg state as a miniature planetary system with the electron orbiting a central core (nucleus plus unexcited electrons) of unit net positive charge. For such a system, the characteristics of the quantum states should be reasonably well described by Kepler's laws.

Kepler's first law, for example, states that the orbit of a planet about the Sun is an ellipse with the Sun at one focus. The electron orbits are also elliptical although for simplicity only circular Bohr orbits are usually discussed in elementary textbooks. In the 'old' quantum mechanics – i.e. the Bohr theory and its various elaborations (principally by Arnold Sommerfeld) predating the creation of a consistent quantum theory in 1925 – the atomic orbits were classified as 'penetrating' or 'nonpenetrating'. The penetrating orbits are highly elliptical (like the orbits of comets) and take the electron near or through the core; the nonpenetrating orbits are more nearly circular (like the planetary orbits of the Solar System) and widely circumnavigate the core (Figure 2.2).

To those who have studied the modern quantum theory of the atom before (if ever) encountering the old quantum theory of electron orbits, the correlation between quantum states and Bohr orbits may at first be a little surprising. For example, the electron probability distribution in an

---

[11] The quantum mechanical expression for the polarisability contains terms involving the product of two radial matrix elements divided by an energy interval. Since a radial matrix element increases as the square of n, and the energy interval decreases as the cube of n, the polarisability increases as the seventh power of n.

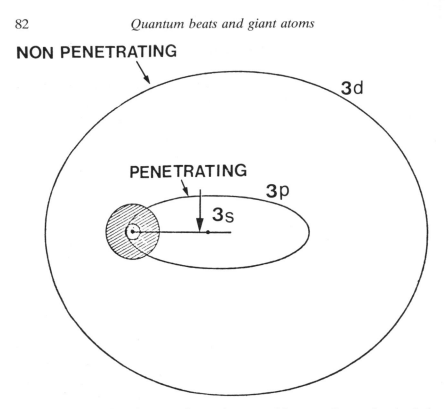

Figure 2.2. The three lowest valence electron orbits according to the classical model of the sodium atom. The shaded circle represents the core of filled electronic shells and the atomic nucleus. Orbits are designated penetrating or nonpenetrating according to whether or not they pass through the core. (Adapted from H. E. White, *Introduction to Atomic Spectra* (McGraw-Hill, New York, 1934) 103.)

S state, a quantum state of zero angular momentum, is spherically symmetric; textbook pictures often represent the S state electron distribution as a fuzzy ball. Classically, however, the state with zero angular momentum is the ultimate penetrating orbit where the ellipse has degenerated into a straight line right through the core. The higher the angular momentum, the more nonpenetrating is the orbit, and the less spherically symmetric is the probability distribution. There is no contradiction here, for what is being pictured are two entirely different things. An orbit represents the sequential motion of the electron in time; a stationary state probability distribution does not represent motion and has no causal implications at all. However, the fact that the S state probability distribution is nonvanishing at the nucleus is consistent with the classical linear trajectory. The quantum wave functions for all other angular momentum states have a node or zero amplitude at the origin.

One might also be surprised to learn how low the angular momentum of a nonpenetrating orbit can be. Consider the orbits corresponding to the lowest three states of the sodium atom, which, like hydrogen, has a single outer valence electron (Figure 2.2). Although the 3s and 3p orbits are clearly penetrating, the 3d orbit remains well outside the core[12]. (Nevertheless, beware of classical pictures! I shall return to this point shortly.)

Kepler's third law states that the square of the orbital period of a planet is proportional to the cube of the semi-major axis (i.e. one-half the long axis) of the ellipse. Thus, it follows for the nearly circular orbits of high angular momentum that the orbital period T_n varies as the $\frac{3}{2}$ power of the radius or, from relation (2.5b), as the cube of the principal quantum number

$$T_n \propto n^3 a_0^{3/2}. \tag{2.6a}$$

According to classical electrodynamics – which should adequately account for radiation production in the domain to which the correspondence principle applies – an oscillating or rotating charged particle should emit electromagnetic waves of the same frequency as the frequency of periodic motion. The Keplerian electron in level n should therefore emit light at a frequency

$$f_n \sim 1/T_n \propto n^{-3} \tag{2.6b}$$

that varies as the inverse third power of the principal quantum number as it continuously spirals inward to a lower energy orbit corresponding to level $n-1$. The above relation readily follows from the quantum mechanical formula, expression (2.1a), for the hydrogen atom energy spectrum; since E_n varies as n^{-2}, the energy interval between levels n and $n-1$, and therefore the radiation frequency, varies as n^{-3} in the limit of large n.

The classical picture of a gentle transition between close-lying nonpenetrating orbits with emission of low-energy radiation is substantiated quantum mechanically by means of the 'selection rules' governing transitions between angular momentum states. The largest value of

---

[12] It is conventional to employ lower-case letters (s, p, d, etc.) for single-electron orbits and upper-case letters (S, P, D, etc.) for the overall quantum state of a multi-electron atom. (For the hydrogen atom there is no distinction.) A good discussion of the orbits of the old quantum theory is given by H. E. White, *Introduction to Atomic Spectra* (McGraw-Hill, New York, 1934), Chapter 7.

angular momentum (in units of $\hbar$) that an electron in level n may have is $n-1$, and, as pointed out before, the emission of a photon carries away one unit of angular momentum. Since a state with angular momentum $n-2$ can occur only in the manifold of states of principal quantum number $n-1$, the emitting electron undergoes a transition from the level n to the level $n-1$ in accord with the classical picture. For electrons with large, but not necessarily maximal, angular momentum there is a range of lower levels that can be reached from a given Rydberg level n. Nevertheless, for n large enough, the energy intervals – and therefore the radiation frequencies – still vary essentially as n^{-3}.

Classical reasoning also allows us to draw an important conclusion concerning the lifetime of the Rydberg states corresponding to nonpenetrating orbits. The total power radiated by a nonrelativistic accelerated charged particle was first shown by the English physicist J. J. Larmor to vary as the square of the acceleration a as follows

$$\text{Radiated Power} = 2e^2a^2/3c^3. \tag{2.7a}$$

This relation is known as the Larmor formula. By Newton's second law of motion the acceleration of the Rydberg electron is proportional to the (inverse square) electrostatic force keeping it in orbit. Thus the radiated power

$$\text{Radiated Power} \propto a^2 \propto r^{-4} \propto n^{-8} \tag{2.7b}$$

varies as the inverse eighth power of the principal quantum number. Since the quantity of energy carried away by each photon varies as n^{-3} (from relation (2.6b)), the time spent in level n

$$t_n \sim (\text{Radiated Energy})/(\text{Radiated Power}) \propto n^5 \tag{2.7c}$$

should vary as the fifth power of n. High angular momentum Rydberg states, then, are predicted to be *very* long lived. This prediction is not inconsistent with the earlier statement that such states are extremely sensitive to environmental perturbations; Rydberg states are long lived when they are left alone.

Unlike the case of a nearly circular orbit, the acceleration of an electron in a penetrating elliptical orbit depends on the electron location. The force – and therefore the acceleration and rate of light emission – are greatest, however, in the vicinity of the perihelion, the point of the orbit closest to the focus where the core is located. The perihelion distance is largely independent of the energy, and therefore of the principal quantum number, of the orbiting particle. Since an electron emits light significantly only when passing through the perihelion, the time spent in orbit is just proportional to the orbital period. Thus, from relation (2.6b)

$$t_n \sim T_n \propto n^3; \qquad (2.7\mathrm{d})$$

the radiative decay lifetime of the low angular momentum Rydberg states should vary as the cube of the principal quantum number. These states, too, are long lived.

According to the Larmor formula, an electron in a penetrating orbit should radiate energy at a greater rate than an electron in a nearly circular nonpenetrating orbit within the same electronic manifold; the acceleration near perihelion, a distance on the order of a few Bohr radii from the core, is much greater than acceleration at a distance of $n^2 a_0$ from the core. This is again substantiated by quantum mechanical selection rules. Low angular momentum states are found in electronic manifolds of both large and low n (provided only that the angular momentum quantum number L does not exceed $n-1$). Since the probability for an electric dipole transition between two states varies as the third power of the frequency of emitted radiation[13], other things being equal, quantum mechanics favours a large 'quantum jump', i.e. a transition from a state n,L to a lower energy state $n',L-1$ with $n' \ll n$. Thus radiative decay of low angular momentum Rydberg states should lead to photons of higher energy than radiative decay from high angular momentum states of corresponding principal quantum number.

The above properties of Rydberg states are reasonably well confirmed experimentally, and one might be tempted to suppose that the simple picture of a distant outer electron orbiting a central nucleus and inner electron core with little mutual interaction is an adequate model for a greatly excited atom – at least for the classically nonpenetrating orbits (which exclude the S and P states). It would then follow that singly excited Rydberg atoms, regardless of the distinguishing properties of the parent ground state atoms, should exhibit essentially hydrogenic behaviour. In many ways this expectation is realised. With regard to binding energies, polarisabilities, lifetimes – and indeed every atomic property of which the calculation involves the radial coordinate r to a non-negative power – a Rydberg atom increasingly resembles a hydrogen atom the larger n becomes.

---

[13] The emitted intensity, proportional to the product of the transition probability and the light frequency, therefore varies as the fourth power of the light frequency in accord with what one would deduce from the classical Larmor formula, relation (2.7a), for an orbiting charged particle.

Nevertheless, upon closer scrutiny, this comfortable agreement crumbles in some rather interesting ways – as in the case of the anomalous fine structure of the sodium atom. Because of the relative simplicity of its electronic configurations, the ease with which one can work with it experimentally, and the convenient region (that of visible light) into which fall many of its spectral lines, the sodium atom makes an excellent system for the investigation of Rydberg states. With a single valence electron outside an inert gas (neon) core, sodium Rydberg states may be expected to resemble closely the excited states of hydrogen. Thus the observation and interpretation of marked nonhydrogenic behaviour of some property of sodium would be of considerable theoretical interest in atomic physics.

To understand what is anomalous about some sodium fine structure levels, let us first reconsider the fine structure of hydrogen, as this is the model for normal structure. I explained previously that the fine structure splitting of the Bohr energy levels originates in the spin–orbit interaction, i.e. the interaction between the electron magnetic dipole moment and the local magnetic field produced by the apparently circulating proton. Since the charge of the electron is negative, the orientation of the electron magnetic moment (proportional to the electron charge) is opposite that of the electron spin. One consequence of this is that the energy of an electron is lowered when the electron magnetic moment is aligned parallel to the local magnetic field; this is the state for which the electron spin and orbital angular momenta are *anti*parallel. Conversely, the electron energy is raised in the opposite configuration where the two momenta are parallel. Thus, the fine structure is considered normal (or hydrogenic) when the state with electron total angular momentum quantum number $J = L + \frac{1}{2}$ lies higher (less tightly bound) than the state with $J = L - \frac{1}{2}$ (Figure 2.3). The energy interval, derived from the Dirac theory of the hydrogen atom, is

$$\Delta E = E_{n,L,J=L+\frac{1}{2}} - E_{n,L,J=L-\frac{1}{2}} = \alpha_{fs}^2 Ry/[n^3 L(L+1)], \quad (2.8)$$

where α_{fs} is again the fine structure constant $e^2/\hbar c \sim 1/137$.

For the D states of excited sodium not only is the magnitude of the fine structure splitting not accurately given by relation (2.8), but the ordering of the levels is reversed, the $J = \frac{5}{2}$ states lying lower than the $J = \frac{3}{2}$ states thereby giving a ΔE of opposite sign. That the sodium D states of the ground ($n = 3$) level are inverted has been known at least since the 1930s by means of optical spectroscopy with high-quality interference gratings. From measurements of the light absorption spectrum

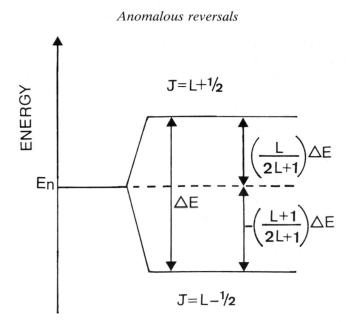

Figure 2.3. Normal (hydrogenic) fine structure level ordering. In the presence of the spin–orbit interaction an energy level E_n ($n > 1$) splits into two levels; the one with electron total angular momentum quantum number $J = L + \frac{1}{2}$ lies above the other with $J = L - \frac{1}{2}$.

corresponding to transitions from 3P to 4D and from 3D to 4P states, experimenters were able to infer the 3D fine structure splitting. Subsequent measurements of the same kind, and by the more recent experimental method known as two-photon spectroscopy[14], showed that the inversion of the sodium fine structure persisted in levels $n = 4, 5$ and 6.

Why is the fine structure in these low-lying levels anomalous? The explanation is not a simple one; there may in fact still be no consensus amongst theoreticians as to the relative importance of various proposed mechanisms. Speaking generally, however, the excited valence electron does not experience a purely central potential ($V \propto 1/r$) as a result of the presence of the other ten electrons that (along with the nucleus) comprise the core. The interaction between the excited electron and the core is referred to as 'core polarisation'. In the classical picture the

[14] Under appropriate conditions, a bound electron can absorb two photons, each of about one-half the energy required to effect the desired transition. By angular momentum conservation the allowed transition is governed by the selection rule, $\Delta L = 0$ or 2 (depending on photon polarisation). Two-photon transitions can occur between the sodium 3S and nD states.

orbits of the core electrons are perturbed by the penetration of the valence electron; this distorts the potential in which the valence electron finds itself and ultimately changes the energies of the two fine structure states from what they would be in hydrogen where there is no subsystem of core electrons. A classical picture, however, can be misleading. Not all sodium fine structure levels are inverted; the $P_{3/2}$ and $P_{1/2}$ levels are normally ordered even though P states correspond to highly penetrating classical orbits.

In accordance with the correspondence principle one might expect that, beyond some threshold value of the principal quantum number, the fine structure ordering must reverse, and the energy splitting become progressively more hydrogenic, as n increases. The spectroscopic method of quantum beats allows this supposition to be tested without at the same time perturbing the states by probing.

Since the experimental task in question requires the measurement of small energy intervals, it is a significant advantage that quantum beat frequencies are insensitive to atomic motion. As is well known from classical physics, the frequency of a light wave (in fact, any kind of wave) emitted by a source moving with respect to an observer is perceived to be shifted, either higher or lower depending on the direction of relative motion, in comparison with the frequency emitted by a stationary source. Known as the Doppler effect, this frequency shift has been the nemesis of many a spectroscopic investigation. Were all atoms to move at the same speed in the same direction, the simple displacement of a spectral line could be taken into account easily. The net effect, however, of a large number of atoms moving in different directions with a wide spread of speeds is to produce 'Doppler broadened' spectral lines whose overlap could obscure fine details of atomic energy level structure.

Since a quantum beat results from the interference between photons that could be emitted from any two of a set of superposed states of the *same* atom, these photons are Doppler shifted to the same extent. Consequently their *difference* frequency, which is the frequency of the quantum beat, is largely independent of the dispersion in atomic velocities[15]. It must be said, of course, that, when the atomic transition actu-

---

[15] The frequency of light emitted by an atom moving at a nonrelativistic speed v towards a stationary observer is of the form $\omega = \omega_0(1 + v/c)$, where ω_0 is the corresponding frequency in the atomic rest frame, and c is the speed of light. The difference in frequency between two waves emitted by the same source is then $\Delta\omega = \Delta\omega_0(1 + v/c)$. Since the difference frequency is many orders of magnitude smaller than the optical

ally occurs, a single atom emits but one photon and does *not* produce a beat in the photodetector output. The observed beat is the product of many such individual emissions – yet the phenomenon does *not* originate in the interference of photons from different emission events – i.e. from different atoms. As in the previously discussed case of interference with free electrons, the *capacity* for quantum interference is intrinsic to each atom. Anyone who finds it hard to visualise just exactly how this occurs is not alone, for this process is again one of the central mysteries of quantum mechanics.

The experiments were performed in the early 1970s at the Ecole Normal Supérieure (ENS) in Paris where I was a guest scientist at the spectroscopy laboratory founded by Alfred Kastler and Jean Brossel. To my good fortune, there had just returned to the ENS a former student, Serge Haroche, who, during a postdoctoral stay at Stanford University in California, had also become interested in quantum beats and was in the process of starting up research in this area at the ENS. We joined forces.

To generate a linear superposition of excited sodium D states directly from the ground state would have required a tunable pulsed laser in the ultraviolet; such a light source did not exist. The problem was solved by exciting the atom in two stages. First, the yellow light from a pulsed dye laser was used to 'pump' sodium atoms from the 3S ground state to the $3P_{3/2}$ state, and then, before the 3P states could decay, the atoms were irradiated with the blue light from another pulsed dye laser to bring them into the desired linear superposition of nD states. Both lasers were tunable; by adjusting the wavelength of the second laser, one could select electronic manifolds of different principal quantum number n.

Since anomalous fine structure had already been observed for levels 3 to 6, we looked for quantum beats in the light issuing from 7D states. There were no beats! Our first thought was that we might have discovered straightaway the 'cross-over' level in which the $D_{5/2}$ and $D_{3/2}$ states are almost degenerate (giving rise to 'beats' of zero frequency). However, the application of a small magnetic field, which reduced the energy interval (and Bohr frequency) between certain $D_{5/2}$ and $D_{3/2}$ states, did generate beats, thereby suggesting that the beat frequency in the absence of a magnetic field was not zero, but rather too large to be produced by our pulsed laser or to be measured by our photodetector.

frequency of either wave, $\Delta\omega_0 \ll \omega_0$, and since $v/c \ll 1$, the dispersion in atomic velocities does not result in any significant broadening of the quantum beat signal.

(By the uncertainty principle, the frequency spread of the second light pulse must be greater than the Bohr frequency associated with the two 7D fine structure levels if a quantum beat is to be produced. Also, the response time of the detector must be shorter than the beat period if the beat is to be detected.) We estimated, then, that the 7D fine structure interval must have exceeded about 150 MHz. This turned out to be the case for 8D states too. Starting with 9D, however, our apparatus began to register field-free quantum beats.

Despite the theoretical simplicity of the experimental procedure, it is worth noting that the experiment had not been an easy one. The two tunable dye lasers, relatively compact and uncomplicated affairs, were themselves pumped by the ultraviolet (UV) radiation from a third laser, a huge and powerful apparatus that shot massive electrical discharges through a chamber of nitrogen gas. The subsequent de-excitation of the nitrogen molecules gave rise to about one million watts of UV radiation. Not only was the electrical noise from these discharges 'deafening' to the rest of the electronic apparatus, but the switching device (or thyrotron), which triggered the release of the large amount of electrical energy stored in an extensive bank of capacitors, often failed to work. Worse still than an electrically noisy laser was a malfunctioning one that sat unproductively quiet. Many laboratory hours were spent in tedious searches through the morass of cables filling the power cabinet of the laser in the (usually vain) hope that the device could be started up again without intervention of the manufacturer's repairman (who was generally servicing another laser somewhere else in Europe).

Nevertheless, with perseverance, the experiment was eventually brought to the state where fine structure quantum beats in a succession of increasingly high Rydberg states could be measured. At that point, unfortunately, my time was up, and I had to leave France. Continuation of the work after my departure led to the surprising result that the quantum beats in levels 9 to 16 showed *no* tendency at all to become hydrogenic. The energy splittings, deducible directly from the beat frequencies, continued to depart from the hydrogenic interval of relation (2.8). And the level ordering remained inverted[16].

The level ordering, it should be noted, is not deducible from the measurement of field-free quantum beats, since the latter provides information only on the magnitude of the energy interval, not on its

[16] C. Fabre, M. Gross and S. Haroche, Determination by Quantum Beat Spectroscopy of Fine-Structure Intervals in a Series of Highly Excited Sodium D States, *Opt. Commun.* **13** (1975) 393.

sign. The order can be determined, however, by a judicious application of the Stark effect, the shifting of atomic energy levels by a static electric field. In the presence of an electric field, all the nD states become more tightly bound, i.e. shift downward on an energy diagram. However, the $D_{5/2}$ substates shift downward to a greater extent than the $D_{3/2}$ substates. Thus, if the $J = \frac{5}{2}$ states lie above the $J = \frac{3}{2}$ states (the normal hydrogenic ordering), the quantum beat frequencies become smaller with increasing electric field (since the lowest $J = \frac{5}{2}$ states approach the $J = \frac{3}{2}$ states). By contrast, if the $J = \frac{5}{2}$ states lie below the $J = \frac{3}{2}$ states (anomalous ordering), the energy intervals – and consequently the quantum beat frequencies – increase with increasing electric field strength. The combined results of previous work and the quantum beat experiments showed that all measured intervals from $n = 3$ to $n = 16$ were anomalous!

Now a principal quantum number of 16 is not exactly close to infinity, which is ideally the limit in which quantum and classical mechanics give equivalent descriptions of a physical system. It is not even close to 500, which represents more or less an upper limit of atomic excitation currently achieved in a terrestrial laboratory. Nevertheless, a sodium atom in the $n = 16$ level is a highly excited atom; it is large enough (classically speaking) to contain over 4000 ground state hydrogen atoms. Moreover, the energy of the excited electron is about 96% of the energy required for ionisation out of the ground state. It was the pattern of measurements, however, that was most significant; the quantum beat frequencies all fell on a smooth empirical curve (Figure 2.4), extrapolation of which did *not* lead to hydrogenic behaviour as n approached infinity.

Why not? Is the correspondence principle violated? Understanding this puzzling experimental result taught me several lessons. The first was not to underestimate what can be learned from the classical model of the atom. Clearly the anomalous behaviour persists because of the interaction of the outer electron with the core. It had been implicitly assumed at the outset that the higher the state of excitation, the less penetrating would be the orbit, and the weaker would be the interaction with the core. We would have done well to think more carefully about the theory of classical orbits. The shape of an elliptical orbit can be quantified by the eccentricity of the ellipse, which is the ratio of the distance of one focus from the centre to the length of the semi-major axis. For example, for a circle both foci coincide at the centre and the eccentricity is zero. The eccentricity of the orbit of an electron in level n with angular

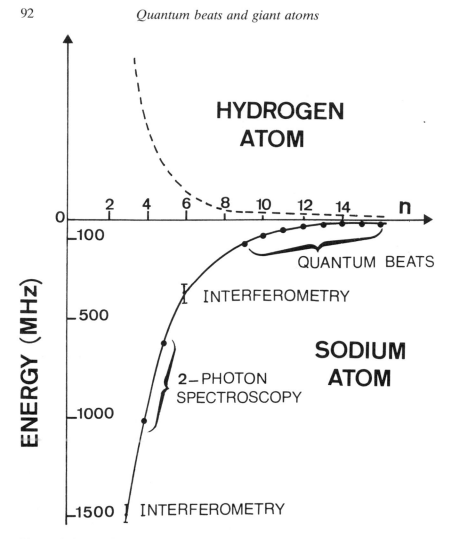

Figure 2.4. nD fine structure intervals of the hydrogen and sodium atoms. Extrapolation of the curves to high n suggests that the sodium fine structure remains anomalous irrespective of the degree of excitation. (Adapted from C. Fabre *et al.*, *Opt. Commun.* **13** (1975) 393.)

momentum quantum number ℓ subject to an inverse square force can be shown to be

$$\varepsilon = [1 - \ell(\ell+1)/n^2]^{\frac{1}{2}}. \tag{2.9}$$

For a state of maximum angular momentum, $\ell = n - 1$, the eccentricity, $\varepsilon = n^{-\frac{1}{2}}$, approaches zero as n approaches infinity, as expected for a circular orbit. However, for the d states ($\ell = 2$) the expression for the

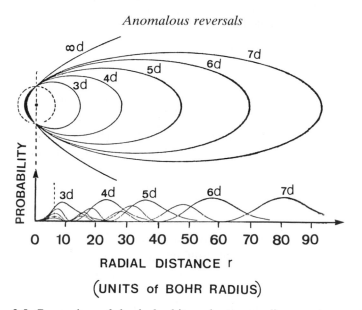

Figure 2.5. Comparison of classical orbits and corresponding quantum mechanical radial probability distributions of the *n*d electron. Although the mean orbital size increases with principal quantum number *n*, the orbital properties near the core are largely insensitive to the degree of excitation. (Adapted from H. E. White, *Introduction to Atomic Spectra* (McGraw-Hill, New York, 1934) 113.)

eccentricity, $\varepsilon = [1 - 6/n^2]^{\frac{1}{2}}$, shows that as *n* increases, the orbit becomes more elongated, and not necessarily less penetrating. In fact, as illustrated in Figure 2.5 for a series of d orbits, at the perihelion the penetration is about the same irrespective of the principal quantum number. Thus, a 16d state might well be expected to interact with the core as much as would a 3d state.

The second lesson was not to overestimate what can be learned from the classical model. The classical 3d orbit was considered to be nonpenetrating and should *not* have interacted significantly with the core. From the perspective of quantum mechanics, however, the radial portion of the *n*d wave function is *always* penetrating – for any *n*. The 'loops' of the wave function extend into the core falling monotonically to zero within a distance of about five Bohr radii from the centre. The spin–orbit interaction depends on the expectation (or mean) value of the inverse third power of the electron radial coordinate (r^{-3}); thus the behaviour of the wave function close to the nucleus and electron core can be significant even if the mean radial distance, $\langle r \rangle \sim n^2 a_0$, is very large.

What kinds of interactions specifically occur between the outer electron and the core to invert the fine structure order? Many studies have been undertaken with varying degrees of success to answer that question. Unfortunately, it would appear that the more quantitatively successful the analysis, the less insightful is its underlying basis in terms of a visualisable mechanism. One of the earliest suggestions is that the inversion results from 'configuration mixing'. The actual state of the excited sodium atom is not exclusively the state to which the valence electron has been nominally excited (e.g. the nd states), but is in fact a linear superposition of other states of excitation, or configurations, brought about by the electrostatic interactions amongst the electrons. One such configuration might include the excitation of the valence electron to the nd state and simultaneous promotion of a core electron to the 3p state. These 'virtual' configurations with two excited electrons cannot be detected directly, but are believed to influence the relative ordering of the nD fine structure levels.

If further experiment and theoretical analysis sustain this picture, then a highly excited atom is indeed a marvellous structure. Nearly macroscopic in size – from the perspective of what can be resolved by a light microscope – it has many of the attributes of a miniature planetary system subject to Kepler's and Newton's laws, while at the same time preserving in the fine details of its energy level structure the effects of strange quantum processes without parallel in the macroscopic world of classical physics.

2.3 Quantum implications of travelling in circles

Looking out over the countryside from the hills of the Hainberg Wald, I could see the red tile roofs of Göttingen rise above the surrounding verdant forest like clusters of mushrooms. Although outwardly similar to other medieval German towns of Lower Saxony, Göttingen was different. At the entrance to the *Rathskeller*, or Town Hall Cellars, the old proverb, *'Extra Gottingam non est vita'*, may once have depicted the world of physics with only mild exaggeration. I went to Göttingen in the late 1970s as a guest professor at the Physikalisches Institut, the institute founded in 1921 by Max Born and James Franck.

To walk through this charming former Hanseatic town largely spared the ravages of time and war, is to walk back in history. All round the town centre are the beautiful old Renaissance frame houses, the Fach-werkhäuser, with white plaster and brown beams, each succeeding level overhanging the street a little further than the previous one. I presume, although I could be mistaken, that this construction afforded the ancient tenants the best configuration for jettisoning their refuse onto the streets below. But the Göttingen I saw was clean and bright.

On Market-place, close by the *alte Rathaus*, or Old Town Hall, stood the *Gänselieselbrunnen*, the Goose-girl Fountain, which was something of a town symbol. Tradition required that male doctoral candidates of the University, dressed in tailcoat and top hat, climb the pedestal and kiss the bronze Goose-girl after passing their examinations. Unfortunately, the *Gänseliesel* had to be removed for repairs, but that minor inconvenience, I soon discovered, did not cause the innovative Göttinger to break with the past. Walking near the fountain one day I encountered a mule-drawn wagon filled with top-hatted *jungen Doktoranden* bringing with them their own quite lively 'Gänseliesel' who took her place on the fountain pedestal. Tradition was preserved!

In the 1920s and 1930s, Göttingen was one of the world's centres of physics and mathematics. It was said to be the 'Mecca of Physics'; David Hilbert, Max Born and James Franck were its 'prophets', and researchers, students and visitors made the pilgrimage there from all parts of the globe. Although this golden age had long passed, I was glad to make my own pilgrimage and draw inspiration from the physical reminders of a period of scientific creativity that was not likely to occur again. I missed that age, having been born too late. To assuage this sense of loss, I often strolled through the quiet *Stadtfriedhof* (cemetery) west of town and looked at the inscriptions on the gravestones which recalled people and events closely associated with the physics and mathematics that intensely interested me. There were Carl Friedrich Gauss, Max Planck, Max von Laue, Max Born, David Hilbert, Otto Hahn and others less well known.

The appointment of Gauss in 1807 as professor of mathematics and director of the *Sternwarte* (Observatory) marked the initial point of ascendancy of Göttingen as a centre of scientific excellence. Gauss was probably the greatest mathematician of his time, but he also employed his extraordinary talents on physically important problems in astronomy, geodesy and electromagnetism. His studies of spatial curvature and non-Euclidean geometry led to mathematical advances that would one day serve Einstein in the creation of general relativity. The tradition of contributing both to pure mathematics and fundamental science passed from Gauss to his former student, Bernhard Riemann, and ultimately to Hermann Minkowski and to David Hilbert, all professors of mathematics at Göttingen. Hilbert's writings on differential equations and eigenvalue problems provided exactly the mathematical foundations needed by the Göttingen quantum physicists – who, to their misfortune, did not pay close enough attention until after Schrödinger 'scooped' them in the discovery of the nonrelativistic quantum mechanical wave equation.

Recorded on the gravestones of a number of those luminaries buried in the *Stadtfriedhof* were words or symbols that distilled from a lifetime of work the core of their goals and achievements. On Hilbert's stone, for example, could be read the words, 'Wir mussen wissen; Wir sollen wissen' ('We must know; we shall know'), epitomising, I presume, his struggle to provide a unified logical foundation to all of mathematics. This dream was effectively shattered by the epoch-making paper of Kurt Gödel on formally undecidable propositions. Max Born's stone bore the basic commutation relation between momentum and coordinate, $pq - qp = (h/2\pi i)\mathbf{1}$, that represented a fundamental distinction between quantum and classical mechanics. He had first written that expression down when he converted into matrix notation[17] a perplexing calculation left with him by his young assistant, Werner Heisenberg. That calculation marked the genesis of a true quantum mechanics. On the tombstone of Otto Hahn was displayed the uranium fission reaction that astounded his contemporaries when they learned that atomic nuclei could be split in half.

Of greater significance to me personally than even the tangible links to the past found in the *Stadtfriedhof* was the intangible, almost spiritual, tie of the Hainberg Wald. How often must those woods have served as the backdrop for intense discussions on quantum physics between the Göttingen physicists and their visitors – between Born, Bohr, Einstein, Heisenberg, Pauli and many others. Heisenberg recalls of his first meeting with Bohr in the summer of 1922, that

... Bohr came to me and suggested that we go for a walk together on the Hainberg outside Göttingen. Of course, I was very willing. That discussion, which took us back and forth over Hainberg's wooded heights, was the first thorough discussion I can remember on the fundamental physical and philosophical problems of modern atomic theory, and it has certainly had a decisive influence on my later career[18].

How I would have liked to hear those seminal conversations whose

---

[17] The coordinate q and linear momentum p are mathematical operators expressible, among other ways, as infinite dimensional matrices. It is from the noncommutativity of operator multiplication that the Heisenberg uncertainty relations formally arise.

[18] B. L. van der Waerden, *Sources of Quantum Mechanics* (North-Holland, Amsterdam, 1967) 22.

echoes faded long ago but reverberate still in the writings of the creators of quantum physics. In an indirect and less momentous way, the Hainberg was to influence my own research.

My wife and I lived on the periphery of Göttingen right across the road from the paths that led into the Hainberg woods. In the mornings I arose before sunrise and worked intensely for hours on a variety of quantum mechanical problems. I would then stop around noontime and go for a run through the woods by a long circuitous loop that took me along tranquil leaf-strewn pathways amongst the tall trees, past fields and farmland and the exercise stations of an *Erholungsgebeit* and eventually back home. Ironically, as I made my way around the Hainberg, I began to think about the curious behaviour of quantum systems that, in a manner of speaking, travel in circles.

The questions that aroused my curiosity at the time concerned the experimental distinguishability of different ensembles of quantum systems. The nature of the problem is subtle; there is no direct parallel in classical physics. If one wants to know whether one collection of macroscopic objects is different from another, he can in principle look at the two collections, count their constituents, probe them, smell them, taste them or whatever. The issue, however, is not so clear when treating a collection of objects whose behaviour is quantum mechanical. According to orthodox quantum theory, specification of the wave function of a system provides all the information about that system that is allowable by physical law. If the system comprises a statistical mixture of subsystems in different quantum states, then the maximum information is contained in the so-called density matrix, effectively a tabulation of the fractional composition of each subset of objects characterised by the same wave function. But is it always clear when two seemingly equivalent wave functions are actually different, or when two seemingly different wave functions are physically the same?

A wave function ψ is not itself an experimentally observable quantity but always enters in bilinear fashion (i.e. in pairs of ψ and ψ^*) the mathematical expressions describing things which are observable. Consequently, the wave function of a quantum system is not unique; for example, the same quantum state can be represented by an infinite number of wave functions differing only by a phase factor of the form $\exp(i\phi)$. In my investigation of the information content and experimental implications of different wave functions, I wondered whether it was truly the case that such a phase factor has no physical consequences at all.

One way by which the wave function of a quantum system can incur a phase factor $\exp(i\phi)$ is through a rotational transformation. The rotational properties of wave functions play an important role in quantum theory; these properties are determined by the angular momentum of the particles whose quantum behaviour the wave functions are presumed to describe. A particle like the electron or neutron, which has an intrinsic spin angular momentum of $\frac{1}{2}\hbar$, is characterised by a *spinor* wave function. A spinor is a mathematical object with two components – like a matrix with two rows and one column. In the context of quantum physics, the upper component gives the 'spin-up' contribution to the wave function while the lower component gives the 'spin-down' contribution. Since the spin attribute of being 'up' or 'down' is defined with respect to an arbitrary quantisation axis, the spinorial components will be modified if a new quantisation axis is chosen. The new components of a spinor are determined from the old components by a rotational transformation.

Suppose the new axis is inclined at an angle θ to the old axis. The rotational transformation peculiar to spinors involves the sines and cosines of $\theta/2$. Now this leads to a curious result, for one can imagine a new axis inclined at 2π radians or 360° to the original axis which, according to our familiar notions of classical reasoning, is no new axis at all; it is the same axis as the original one. Nevertheless, the algebraic prescription governing the rotation of spinors gives rise to the transformation

$$(\text{new spinor}) = e^{i\pi}(\text{old spinor}) = -(\text{old spinor}),$$

that is, each component of the new spinor is the *negative* of the corresponding component of the old spinor. The negative sign can be regarded as a phase shift of π radians, since $e^{i\pi} = -1$ (an expression that was, itself, at one time rather puzzling[19]). A rotation by 360° does not reproduce the same spinor wave function.

In so far as one is discussing abstract mathematical objects and theoretical changes of coordinate axes, the above rotational property of spinors need not be disturbing. Purely mathematical relations do not have to satisfy criteria imposed by the real world. But physics obviously must. According to quantum mechanics the 'passive' view of a rotation

---

[19] In general, $e^{i\phi} = \cos\phi + i\sin\phi$. The identity, $e^{i\pi} + 1 = 0$, discovered by Leonhard Euler (1707–83), contains the most important symbols of modern mathematics and has been regarded in the past as a sort of 'mystic union' in which 0 and 1 connoted arithmetic, π stood for geometry, $i = (-1)^{1/2}$ designated algebra, and the transcendental e represented analysis. It seems fitting, somehow, that this identity be the point of origin of another of the 'mysteries' of quantum mechanics.

as a reorientation of the coordinate axes with the wave function (e.g. a spinor) held fixed is equivalent to the 'active' view that the wave function itself is rotated with respect to a fixed coordinate system. If spinors are to be suitable representations of actual fermionic particles, it is then a legitimate question to ask what, if any, are the observable consequences of physically rotating by 360° a system characterised by a spinor.

The inferences to be drawn from quantum mechanics texts and monographs seemed to indicate that *no* consequences would result. The 'holy P.A.M. himself', as Schrödinger referred to Dirac, asserted in his classic work, *The Principles of Quantum Mechanics*[20]

We thus get the general result, *the application of one revolution about any axis leaves a ket unchanged or changes its sign according to whether it belongs to eigenvalues ... which are integral or half odd integral multiples of ħ.* A state, of course, is always unaffected by a change of sign of the ket corresponding to it. [Italics included in the original text.]

(The idea of bras and kets, it should be noted, was created by Dirac to represent the state of a quantum system in a totally general way; when a ket is combined in a specified way with a coordinate bra, the resulting bracket (bra-ket) yields the wave function.) Dirac's general result has been frequently cited in the pedagogical literature of quantum mechanics; since the results of all measurements are representable by expectation values (in effect, integrals) bilinear in the wave function, two wave functions differing only in overall sign cannot lead to different physical predictions.

Although the above conclusion is not incorrect, it nevertheless seemed to me that its application to particle rotation entailed further thought. For one thing, I discerned a distinction between 'mathematical' rotations (whether viewed passively or actively) that involved the reorientation of a coordinate system or mathematical function like the spinor wave function, and a 'physical' rotation whereby some force or interaction was required to cause a real particle such as an electron to depart from rectilinear motion. The former case, whereby an isolated system (measuring apparatus included) is rotated, has no experimental counterpart, and the global phase factor to which it gives rise is, as Dirac states, not observable. But this is not what is ordinarily meant by a rotation. It is the latter case where only a part of the system is rotated

---

[20] P. A. M. Dirac, *The Principles of Quantum Mechanics*, 4th Ed. (Oxford, London, 1958) 148.

relative to a fixed part that provides a reference against which the rotation is measured. The 2π phase change of a spinor-characterised portion with respect to a fixed portion of a larger encompassing system *does* have physical implications.

There is an interesting example drawn from classical physical optics that illustrates in an analogous way the issues underlying the observability of spinor phase. In the theory of scalar diffraction – which ignores the electromagnetic nature of light and the corresponding property of light polarisation – the amplitude of diffracted light at some point P is given by the so-called Helmholtz–Kirchhoff integral. The precise specification of the light amplitude $\psi(P)$, which can be found in almost any optics textbook, is not needed here, but it will be instructive to note that it takes the form

$$\psi(P) = -\,i \times \{\text{integral over diffracting surface}\}.$$

Textbooks discuss nearly all aspects of this amplitude that derive from the surface integral: the dependence on wavelength, the angle of diffraction, the approximations that lead from Fresnel (near-field) to Fraunhofer (far-field) diffraction, and the application of the integral to various obstacles and apertures. Yet I am aware of no standard text that considers the experimental consequences of the factor $-i$. Although it corresponds to a phase shift of 90° or $\pi/2$ radians [$\exp(-i\pi/2) = -i$] between the incident and the diffracted light, a reader may well be left with the feeling that it is an artefact of the mathematical approximations underlying the derivation and therefore of no physical significance. But this is not so. By superposing, as in holography, a coherent background of incident light on the radiation diffracted by an object, one can observe in the resulting interference pattern the effect of the $\pi/2$ phase shift[21].

The above parallel with light suggests that the 360° rotation of an object characterised by a spinor wave function should be observable as well by means of some type of *quantum* interference. And this is indeed the case. Split-beam interference experiments with neutrons[22], in which the spin of a neutron in one component of the beam was made to precess in a magnetic field relative to the spins in the field-free component,

[21] I first learned of such an experiment by Brian Thompson of the University of Rochester, New York, while participating in his 1977 National Science Foundation Short Course on Coherent Optics.

[22] H. Rauch *et al.*, Verification of Coherent Spinor Rotation of Fermions, *Phys. Lett.* **54A** (1975) 425; S. A. Werner *et al.*, Observation of the Phase Shift of a Neutron Due to Precession in a Magnetic Field, *Phys. Rev. Lett.* **35** (1975) 1053.

showed that the intensity of the recombined beam oscillated as a function of the magnetic field with a periodicity indicative of a spin rotation of 720° (or 4π radians), rather than 360°.

The interpretation of the double-beam interference experiments on neutrons, however, is not without its difficulties. At the core of the problem is the Heisenberg uncertainty principle: one can never know whether a particular neutron followed a path through the magnetic field or through the field-free region. Although in principle the relative spin rotation of neutrons following two classical paths can be measured, this measurement would destroy the interference pattern. Since simultaneous observations of neutron spin rotation and of the interference pattern are incompatible, the notion of relative rotation ceases to have a meaning as it corresponds to nothing which is measurable – at least in the case of fermions.

The problem does not arise, however, for massive bosons[23] because there is always an additional quantum state (the $M = 0$ state) that is insensitive to the presence of the magnetic field. (The photon, though often said to be a spin-1 particle, is excluded for it has only two spin components[24].) The theoretical expression for the interference pattern may then be decomposed into an incoherent sum of two terms: one characterising the initial quantisation axis of the particles, the other representative of a rotation with respect to that axis. For bosons, therefore, the concept of relative spin rotation retains a meaning in the setting of a split-beam interference experiment. However, when a boson wave function is rotated by 360°, nothing 'interesting' happens, for it leads to the same wave function. Hence, there is not much point in studying rotated bosons.

To avoid ambiguities, either semantic or otherwise, associated with the question of spinor rotations, I thought about the physical implications of wave function transformations in a more general context. The idea of whether *cyclic* quantum transitions were detectable first occurred to me in the form of a mental game as I jogged over the Hainberg. I imagined looking at a friend who is sitting in a chair facing

[23] J. Byrne, Young's Double Beam Interference Experiment with Spinor and Vector Waves, *Nature* **275** (1978) 188.
[24] The spin components of the photon are oriented either parallel or antiparallel to the direction of propagation; a transverse ($M = 0$) substate is disallowed as a result of the vanishing photon rest mass. This profound point is discussed by E. P. Wigner, Relativistic Invariance and Quantum Phenomena, in *Symmetries and Reflections* (Indiana University Press, Bloomington, 1967), Chapter 5. It is more accurate to attribute to the photon a 'helicity' (projection of spin onto linear momentum) of $1\hbar$.

me. I turn my back to him, and he can either do nothing or get up, leave the room and return to his seat. When I turn around, he is in the chair facing me as before. Can I tell by looking at him whether or not he had left his seat?

In the quantum world of atoms this question has a strange and interesting counterpart. Is there a physically observable difference, for example, between an atom that has undergone a transition to a different state and then returned to its original state, and an atom that has never left its original state at all?

According to quantum mechanics, the properties of an atom ought to depend only on its current state, not on its past history. If an atom is in a 3S state, then it manifests all the properties expected of the 3S state irrespective of how it happened to get there – i.e. whether it was produced by decay from a 3P state, decay from a 5F state or absorption from a 2P state. In general, all atoms in a particular quantum state are indistinguishable. Nevertheless, the idea of a cyclic transition has experimental implications. One day, while running and thinking about this question, there occurred to me the possibility of an experimental demonstration by means of quantum beats.

Let us consider an atom (practically speaking, a collection of many identical atoms) with *three* nondegenerate excited states; a pulse of light drives the atom into a linear superposition of the lower two (Figure 2.6). Observed as a function of time following the excitation, the fluorescent light intensity of a specific polarisation will oscillate in the familiar way at an angular frequency ω_{12} corresponding to the energy interval of the two superposed states

$$I_0(t) = A + B\cos(\omega_{12}t + \phi). \tag{2.10a}$$

Here A, B and ϕ are again constants that depend on the electric dipole matrix elements for excitation and spontaneous emission. (I ignore, as inessential in the current discussion, the exponential decay factor.) Suppose, however, that to this standard quantum beat experiment one adds a radiofrequency (rf) electric field that can induce transitions between the excited atomic states 2 and 3. In general, after exposure to the rf field, the atom is in a linear superposition of all three excited states.

When the condition of resonance is met, whereby the radiofrequency exactly matches the Bohr frequency ω_{23}, the theoretical description of the effect of the rf field simplifies nicely. It precisely resembles, in fact, the mathematical expression for rotation of a spinor; the rotation angle corresponds effectively to twice the product of the rf transition matrix

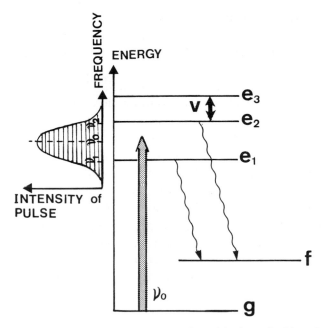

Figure 2.6. Energy level diagram of an atom impulsively excited by a laser pulse (of mean frequency ν_0) from its ground state g into a linear superposition of states e_1 and e_2 and possibly made to undergo a cyclic transition between excited states e_2 and e_3 by an external perturbation V. Whether or not the cyclic transition had occurred can be inferred from the ensuing quantum beat signal in the spontaneous emission to lower state f.

element and the time of exposure to the field. Setting the rf field strength or exposure time so that the corresponding rotation angle is π radians results in all the atoms in state 2 being driven into state 3. A 'rotation' of 2π results in all atoms in state 3 being driven back again to state 2. However, like a 360° spinor rotation, in the process of leaving and returning to state 2, the amplitude incurs a phase shift of π radians; it is the negative of what it was before the transition.

The occurrence of the minus sign would not ordinarily be observed in the spontaneous emission from an atomic state, since the transition probability depends on the square of the absolute magnitude of the amplitude. It is to be recalled, however, that the transitions engendered by the rf field have not, under the conditions of the experiment, affected the amplitude for remaining in excited state 1. The amplitude for excited state 1 is analogous to the coherent background radiation in a holographic demonstration of the $-i$ phase factor. Thus, the negative sign of the amplitude for state 2 represents not a global phase factor, but rather a 180° phase shift relative to the amplitude of state 1. An experimental consequence of this phase shift shows up as a reversal in sign of the oscillatory component of the fluorescent light intensity

$$I_{2\pi}(t) = A - B\cos(\omega_{12}t+\phi). \tag{2.10b}$$

One can enhance the effect by measuring the difference signal

$$I_0(t) - I_{2\pi}(t) = 2B\cos(\omega_{12}t+\phi) \tag{2.10c}$$

and thereby eliminate the constant term. Without the spinorial sign change, the difference signal would be identically zero.

For an atom in a linear superposition of eigenstates, there *can* be an experimentally observable distinction between 'doing nothing' and undergoing a cyclic transition from one of the component states. This does not contradict Dirac's assertion regarding the nonobservability of global phases, but rather serves to emphasise that, whether a phase is global or not, depends on what one does to, or with, the system. The $(-i)$ phase factor, for example, in the diffraction of light is an unimportant global phase factor if one simply measures the intensity of the diffracted radiation; it becomes a relative phase, however, in the holographic recording of this scattered light. The history of a system does matter.

* * * * * * * * * * * *

The strange nature of cyclic transformations in quantum physics impressed itself upon me once again not long afterwards when I was studying the interaction of charged particles with inaccessible magnetic fields, such as occur in the Aharonov–Bohm (AB) effect. The various experimental configurations I described in Chapter 1 all involve the propagation of *unbound* charged particles, i.e. particles that leave their source, diffract around a current-carrying solenoid (or similar structure) and are detected. My focus on atoms made me wonder about an entirely different type of AB configuration – one like a giant planetary atom in which an electron orbited, not a nucleus, but a long solenoid confining a magnetic field. What effect would the magnetic field have on the orbital properties of the electron?

Like other problems relating to the AB effect, this one, too, has its subtleties. The wave function $\psi_M(2\pi)$ of a particle with well-defined angular momentum M (in units of $\hbar$) that has wound once around the cylinder is simply related to the initial wave function $\psi(0)$ by a phase factor as follows

$$\psi_M(2\pi) = \exp(2\pi i M)\psi_M(0). \tag{2.11a}$$

To determine whether or not the 'rotated' wave function is equivalent to the initial wave function requires that one knows the values that the number M is allowed to take.

In the absence of the inaccessible magnetic field, the problem reduces to the quantum mechanical two-dimensional, or planar, rotator, a familiar system whose properties are well understood. The spectrum of angular momentum eigenvalues is the set of all integers, $M = 0, \pm 1, \pm 2$, etc., where states with angular momentum quantum numbers differing only in sign correspond to circulations about the origin in opposite directions. As in the case of the corresponding classical system, the kinetic energy of rotation is proportional to the square of the angular momentum. Thus, pairs of states with angular momenta $\pm|M|\hbar$ are degenerate – as one would infer from the symmetry of the system; there is no reason to expect that, in the absence of external forces, a clockwise rotating particle should have a different energy than one rotating counterclockwise.

With the confined magnetic field present, however, analysis of the planar rotator leads to a curious ambiguity, for two entirely different solutions to the equation of motion (Schrödinger equation) emerge. According to one solution, the magnetic field has no effect on the energy of the system, but leads to angular momentum eigenvalues that depend on the magnetic flux Φ

$$M = M_0 + \Phi/\Phi_0, \tag{2.11b}$$

where M_0 is an integer (one of the eigenvalues of the field-free planar rotator) and Φ_0 is the value of the fluxon

$$\Phi_0 = hc/e. \tag{2.11c}$$

According to the other solution, however, the magnetic field has no effect on the angular momentum eigenvalues, but leads instead to system energies that depend on the magnetic flux. Which solution, then, gives the 'right' answer?

It should be noted here, because the distinction is now important, that the angular momentum that enters the phase factor of relation (2.11a) is the *canonical* angular momentum. This is the dynamical variable that determines (through the commutation relations of its components[25]) the

---

[25] The commutation relations of the canonical angular momentum take the form: $(L_x L_y - L_y L_x) = i\hbar L_z$ with corresponding forms for even permutations of the coordinates x, y, z.

behaviour of a physical system under rotation. It is *not* necessarily the same thing as the 'quantity of rotational motion' which, for the circular trajectory of a point particle, is familiarly given by (mass) × (speed) × (orbital radius). This latter dynamical variable is the *kinetic* angular momentum; it is always an observable quantity, whereas the canonical angular momentum need not be. For the field-free planar rotator there is no difference between the kinetic and canonical angular momenta. When the rotator is in the presence of a vector potential field, however, these two dynamical quantities are no longer the same.

Careful examination of the origin of the two solutions shows that one is not 'more correct' than the other, but rather that they refer to physically different systems, and, as in the case of cyclic atomic transitions, the history of the system plays a significant role. The quantum states of the second solution, in which the energy is flux-dependent, characterise a particle orbiting a confined magnetic field that was 'turned on' at some indefinite time in the past. By Faraday's law of induction (one of the Maxwell equations of classical electromagnetism), an electric field is produced throughout the time interval that the magnetic field is growing from its initially null value to the final constant value it will subsequently maintain. This electric field exerts a torque on the particle thereby doing work and changing the initial kinetic angular momentum and kinetic energy of the particle to the values that characterise the second solution. By contrast, the solution with energy independent of flux represents a system in which the particle orbits a region containing an already existing uniform magnetic field. Such a field does no work on a charged particle (even if the particle were immersed in the field[26]) and therefore cannot alter the particle energy and kinetic angular momentum. Nevertheless, as we have seen before, a constant magnetic field can have quantum mechanical consequences with no counterpart in classical physics. In the present case, the bound-state AB effect modifies the spectrum of canonical angular momentum eigenvalues, and *this* has implications for cyclic transformations.

From relations (2.11a,b,c) it is seen that the wave function of a particle that has undergone N revolutions about an inaccessible constant magnetic field is given by

$$\psi(2\pi N) = \exp(2\pi i N \Phi / \Phi_0)\psi(0). \tag{2.11d}$$

---

[26] A uniform time-independent magnetic field produces a local Lorentz force that acts at right angles to the direction of particle motion. Since force is perpendicular to displacement, the force can do no work on the particle.

Is there any physical distinction between orbiting the magnetic field once and making multiple passages around the field? If the magnetic flux is an integral multiple of the fluxon, the cyclical rotations will have no effect on the particle wave function for any phase factor of the form $\exp(2\pi i \times \text{integer})$ is unity. There is no general reason, however, for the flux to be quantised, in which case the ratio Φ/Φ_0 can be arbitrary. If a value of Φ is chosen such that the phase in relation (2.11d) is π radians for one rotation, then the wave function changes sign for an odd number of rotations just like a spinor. This is very interesting, because nothing has heretofore been said about the spin of the orbiting particle. It could, in fact, be a spinless particle – a charged boson – which, as a result of the presence of the magnetic flux behaves like a fermion under rotation! For arbitrary values of Φ/Φ_0, the wave function characterises a particle that behaves under rotation neither as a fermion nor as a boson. Some physicists today call it an 'anyon'.

How might such strange behaviour be manifested? One possibility is to employ again a split-beam quantum interference experiment. Imagine dividing a collimated beam of charged particles coherently into two components, one of which is made to circulate N times in a clockwise sense about an AB solenoid bearing flux Φ_1, while the other circulates an equal number of times in a counterclockwise sense about a second AB solenoid bearing flux Φ_2 (Figure 2.7). Suppose that the magnetic fields within the two solenoids are oppositely directed. Suppose, too, that the 'bending' magnetic fields outside the cylinders are uniform and equal in magnitude so that the radius of the particle orbit around each solenoid is the same. With this configuration there is no net contribution to the relative rotational phase shift from the external magnetic fields or from the spin angular momentum of the particle. Upon recombination of the components at a distant detector, the forward beam intensity can be shown to vary with the magnetic flux within the solenoids as

$$I(2\pi N) \sim I_0 \cos^2\{N(\Phi_1 - \Phi_2)/\Phi_0\}, \tag{2.12}$$

where I_0 is the incident beam intensity.

The magnetic flux dependence of the signal reveals the topological parameter N, known as the winding number. In a two-dimensional space there is a topological distinction between closed paths that make an unequal number of turns about the symmetry axis. Two such paths cannot be converted into each other by a continuous deformation without being cut. In the split-beam AB experiments described in the

DETECTOR

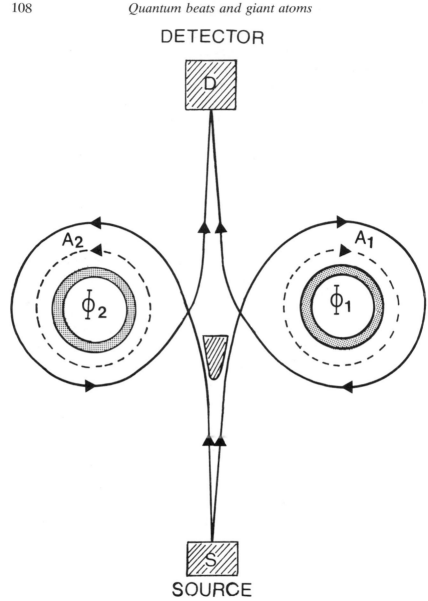

SOURCE

Figure 2.7. Schematic diagram of a split-beam electron interference experiment whereby an electron, issuing from source S, makes an integral number of revolutions about one or the other flux-bearing solenoid. Interference in the forward intensity of the recombined beam reveals the number of 'windings'. For appropriate settings of the magnetic flux, charged bosons can be made to behave like fermions under rotation.

previous chapter, the charged particles propagate from source to detector by two types of topologically different paths: those that pass once to the left side of the solenoid, and those that pass once to the right. Some theoretical work suggests that a complete description of the AB effect should take account of contributions from paths of all possible winding numbers connecting the source and the detector. Thus, besides the configuration of two standard classical paths, one would need to include configurations where either the left or the right or both paths make one or more full loops around the solenoid before extending to the detector. On the basis of the experiments that have been done so far, it would seem that such a description may not be needed; the observed fringe patterns can be adequately accounted for by assuming that only the two classical paths contribute. The proposed experiment, were it to confirm the prediction of relation (2.12), would provide unambiguous evidence of the influence of winding numbers in the Aharonov–Bohm effect.

* * * * * * * * * * * *

In the years following my stay in Göttingen, quantum systems made to undergo some kind of cyclic process have become of widespread interest in physics. Pursuant to the work of M. V. Berry[27] in particular, deep and beautiful connections have been shown between such seemingly disparate concepts as quantum phase and spatial curvature, and amongst phenomena as diverse as spinor rotation, optical activity (the rotation of the electromagnetic field of linearly polarised light), the Aharonov–Bohm effect and superconductivity.

Like all cyclic paths, my own in Göttingen eventually drew towards a close as the day came for me to return home. Setting out for the train station very early on a cold, wintry morning I took one last look at the countryside from which I derived so much pleasure. It was snowing, and in the bright moonlight the Hainberg was radiant with a thick, soft, white mantle of snow. Tall fir trees, their pinnacles lost in the blackness of the sky and their heavily laden branches slung low, lined the path from my front door to the roadway like sentinels before a magical forest. Elves could have walked out of the woods at that moment, and I would not have been surprised, so vividly did the scene recall the enchanting landscapes of the old German *Märchen*. As the sight of woods gave way to the sight of houses, the sense of ancient mystery faded, and an unforgettable experience came to an end – as had that

[27] M. V. Berry, Quantal Phase Factors Accompanying Adiabatic Changes, *Proc. R. Soc. Lond.* A**392** (1984) 45.

marvellous period over sixty years ago when quantum mechanics was created in Göttingen.

2.4　Long-distance beats

As I explained earlier, quantum beats are produced by the radiative decay of *individual* atoms (or molecules) excited into a linear superposition of nondegenerate energy eigenstates. The fact that each atom may be in a superposition state does not, in itself, guarantee that the ensemble as a whole will manifest quantum interference effects, since a large dispersion in the relative phases of the wave function from atom to atom will lead to unsynchronised emissions and hence to no net modulation of the light intensity. In some seemingly paradoxical cases – for example, the 'restoration' of beats by a sufficiently intense broadband laser pulse of *long* duration – the unsynchronised excitation of atoms need not, in fact, lead to a dispersion in phase large enough to destroy the beats. Nevertheless, if there is one thing that might seem certain – for a science whose foundation (metaphorically speaking) rests on the uncertainty principle – it should be this: an atom that is *not* in a linear superposition of its energy eigenstates cannot give rise to quantum beats. Right?

Well, not exactly ... even though to believe otherwise may seem patently absurd and a violation of fundamental quantum mechanical principles. Indeed, so counterintuitive was the experimental possibility that presented itself to me one day when I was thinking about the Einstein–Podolsky–Rosen paper, that despite long acclimatisation to the intricacies of quantum physics I was, myself, startled by its strange implications. I think of this effect, which accentuates the intrinsically nonlocal features of quantum mechanics, as 'long-distance beats'.

It should be mentioned at the outset that the production of quantum beats is not restricted to single-atom systems. Indeed, under appropriate circumstances, the modulation of atomic fluorescence can also occur as a result of the linear superposition of states of a multi-atom system. Consider, for example, two identical atoms, each with a single ground state g and two close-lying excited states e_1 and e_2. If the atoms are near enough to one another – i.e. separated by a distance shorter than an optical wavelength – an incoming photon could be absorbed by either atom and raise that atom to one or the other of its excited states. Now if the excited atom radiatively decays back to its ground state, the final condition of the system is simply two ground state atoms and an emitted photon.

This may be summarised in the following way:

Process I

Atom A: $g \rightarrow e_1 \rightarrow g + \text{photon } \omega_1$

Atom B: $g \longrightarrow g$

(2.13a)

Process II

Atom A: $g \longrightarrow g$

Atom B: $g \rightarrow e_2 \rightarrow g + \text{photon } \omega_2$,

where each photon is designated by its angular frequency (a measure of its energy). The frequency of the emitted photon depends on the excited state from which emission occurs, but, if the detection process does not discriminate between photons, then there is no way to tell which atom had been excited. Consequently, Processes I and II are indistinguishable; to determine the net probability of photon emission, one must add the *amplitudes* of the two processes. The result is that the light intensity, to which photons emitted by one or the other of the paired atoms contribute, is modulated at the Bohr frequency corresponding to the energy interval of the excited states, $\omega_{12} = \omega_2 - \omega_1$.

It is important to note that the excited atom *must* decay back to its ground state if quantum beats are to occur. Were it to decay to some other low-lying state – call it f, for example – then the above two processes (with g+photon replaced by f+photon) would be distinguishable, because one could in principle search out the atom in state f and thereby determine which atom had been excited. For processes with distinguishable outcomes one adds probabilities, not amplitudes; no quantum interference then occurs.

Since the two atoms of the pair may have different velocities relative to a stationary observer, the photon emitted from one atom can be Doppler shifted to an extent different from that of the photon emissible by the other atom. (Remember that only *one* photon is actually emitted by a pair of atoms. The beat arises not from the interference of two simultaneously present photons, but from the interference of probability amplitudes describing the two radiative processes that could potentially occur.) Depending on the distribution of atomic velocities, the spread in photon frequency can be much greater than the Bohr frequency with the result that quantum beats from different atoms would be out of phase; no net modulation of the atomic fluorescence would be observable. As pointed out previously, the quantum beats produced from single-atom systems are not sensitive to the Doppler effect.

An interesting alternative to the sequence of events (2.13a) is one in which either atom is brought into the *same* excited state, for example e_1. In fact, one could dispense entirely with the need for two excited states, and quantum beats could still occur. Experimental evidence for just such an effect has been provided by the photodissociation of diatomic calcium molecules[28]. An incoming photon dissociates the diatomic molecule into two atoms, either of which could be raised to an excited state and subsequently decay emitting a photon different in frequency from the one that was absorbed. The two interfering pathways may be outlined as follows:

Process I :   Incoming photon $+ M_2 \rightarrow M + M^* \rightarrow 2M +$ photon ω_0,

$$(2.13b)$$

Process II: Incoming photon $+ M_2 \rightarrow M^* + M \rightarrow 2M +$ photon ω_0,

where M_2 represents the diatomic molecule, M a ground state atom, M^* an excited atom and ω_0 the photon angular frequency in the rest frame of the emitting atom. If there is no excited state energy interval, then what determines the beat frequency? Upon dissociation, the two atoms recoil with equal, but oppositely directed, velocities of magnitude v. Thus, with respect to a stationary observer in the laboratory, the frequency of the photon emitted in Process I is Doppler shifted in the opposite direction to that of the photon emitted in Process II. The quantum amplitudes for emission of the differentially Doppler-shifted photons interfere giving rise to beats at the frequency $2(v/c)\omega_0 \cos\theta$, where θ is the inclination of the axis of the dissociating molecule to the observation direction. The Doppler effect may make the light beats from processes (2.13a) impossible to observe, but it is essential for the production of quantum beats by processes (2.13b).

One conceptually important feature of the single-atom quantum beat phenomenon is that the beat frequencies, according to standard quantum theory, always correspond to level splittings of the emitting upper states and *never* to level splittings in the final lower states. Indeed, this feature has served as one of the tests distinguishing quantum electrodynamics from competing theories of radiative phenomena based on semiclassical considerations. The reason that beats at the Bohr frequencies of the final states cannot occur is that the decay pathways to

[28] P. Grangier, A. Aspect and J. Vigue, Quantum Interference Effect for Two Atoms Radiating a Single Photon, *Phys. Rev. Lett.* **54** (1985) 418.

alternative final states are distinguishable, and therefore the amplitudes for these processes cannot interfere with one another.

Interestingly, in a two-atom system the quantum beats *can* occur at frequencies corresponding to *final* state splittings. Consider, for example, an ensemble of atoms each with two nondegenerate ground states g_1 and g_2, and one excited state e. Suppose that two atoms, one in state g_1 and the other in state g_2, are irradiated with light of spectral width greater than the ground state Bohr frequency. An incoming photon, then, could excite either one atom or the other to the state e from which the atom subsequently decays by emission of a photon. If it is once more required that each atom radiatively decays back to its original state, then the following two processes are indistinguishable

$$\text{Process I} \qquad \begin{array}{ll} \text{Atom A:} & g_1 \rightarrow e \rightarrow g_1 + \text{photon } \omega_1 \\ \\ \text{Atom B:} & g_2 \longrightarrow g_2 \end{array}$$

$$(2.14)$$

$$\text{Process II} \qquad \begin{array}{ll} \text{Atom A:} & g_1 \longrightarrow g_1 \\ \\ \text{Atom B:} & g_2 \rightarrow e \rightarrow g_2 + \text{photon } \omega_2 \end{array}$$

to the extent that the detector again does not discriminate between the energy of the emitted photons. Interference between the amplitudes for these indistinguishable processes leads to a quantum beat at the frequency $|\omega_2 - \omega_1|$ corresponding to the ground state Bohr frequency.

It is implicit in all the foregoing discussion that, in order for quantum beats to be observable in either the initial upper or final lower states of a radiative transition, the two atoms must be within an optical wavelength of each other. One might think that this limitation on atomic separation is attributable exclusively to the excitation process. That is, if the atoms were further apart than an optical wavelength, then an incoming photon could not 'reach' both atoms simultaneously with the result that there would then not be two indistinguishable excitation pathways. This is not strictly the case, for the spatial extension of a wave packet – which provides a more appropriate description of a photon than does a plane wave – can be much longer than the mean wavelength (as I discussed in Chapter 1 in the context of the wave-like properties of free electrons). The limitation is actually connected with the decay process.

A photon is not detected instantly after emission, but only after the so-called retarded time interval r/c, where r is the distance between the de-excited atom and the detector. The amplitude for the emission and detection of a photon of angular frequency ω (wavelength $\lambda = 2\pi c/\omega$) contains the phase factor $\exp(i\omega r/c)$. Ordinarily, this phase factor does not play a significant role, for it vanishes in the calculation of the corresponding probability. However, when the photon can be emitted by either of two atoms, A or B, then a phase factor containing the appropriate atom-detector distance r_A or r_B appears in the amplitude for each indistinguishable pathway. The emission probability will then vary harmonically (i.e. in an oscillatory way) with the phase $(2\pi/\lambda)|r_A - r_B|$. If the separation between the atoms is much larger than a wavelength, the preceding phase will vary rapidly with the point on the detecting surface at which the photon is received, and the quantum beat signal will be effectively averaged away.

Although the foregoing two-atom quantum beat phenomena have their points of interest, they do not, I think, present conceptual difficulties beyond those already intrinsic to the interpretation of quantum beats from single-atom systems. Two closely separated atoms may be regarded more or less as a kind of 'bondless' molecule, and, after all, the quantum interference of molecular states is no more problematical than the interference of atomic states. To put the matter a little differently, an experimentalist who has assembled the apparatus necessary for coherently exciting individual atoms would not be too surprised to find that he has also coherently excited pairs of atoms if the atomic density were high enough (on the order of 1 atom per cubic wavelength). All the atoms are still close together in one small container and subjected to the same beam of light.

But is it conceivable that two atoms – one in London and the other in New York, for example – can be coherently excited, and yet the local experimentalist making observations in each respective laboratory would not even know? Let us return to the problem that intrigued me.

Imagine a transparent container (a resonance cell) filled with an atomic vapour excited by pulses of light. Each atom has a ground state g, nondegenerate excited states e_1 and e_2, and some lower state f (not necessarily the ground state) to which the excited states can radiatively decay. Disregarding for the moment the precise nature of the light source, I will simply say that photons arrive regularly at time intervals longer than the excited state lifetimes (so that the interaction of an *excited* atom with an incoming photon and the possibility of stimulated emission can be ignored) and are distributed randomly over two possible frequencies. If an arriving photon has frequency ω_1, an absorbing atom is raised from the ground state to excited state e_1. Similarly, the absorption of a photon of frequency ω_2 raises the atom to excited state e_2. Note, however, that the two kinds of photons are each sufficiently sharply defined in frequency so that *no* one photon can raise an atom into a linear superposition of excited states e_1 and e_2. This is important because, although a local observer may not know what kind of photon is going to arrive, he does know – or at least he thinks he knows! – that each atom is excited into a well-defined energy state and not a superposition of states. The seminal condition (2.4b) for the occurrence of quantum beats is not met, and consequently the fluorescent light emission following each excitation should simply decay exponentially in time[29].

In fact, no measurement – light intensity or otherwise – made only on the sample of atoms in the cell would cause the observer to think that the atoms were in some way coherently excited and capable of exhibiting quantum interference. This is rigorously demonstrable by examining the quantum mechanical density matrix for the atoms, i.e. the mathematical construction that provides in principle a complete theoretical description of the states of the atoms. For example, if the atoms in the cell were excited into linear superpositions of states ψ_1 and ψ_2 representable by a wave function of the form

$$\psi = a_1\psi_1 + a_2\psi_2, \tag{2.15a}$$

then the density matrix $\boldsymbol{\rho}$ would take the form

$$\boldsymbol{\rho} = \begin{pmatrix} \langle |a_1|^2 \rangle & \langle a_1 a_2^* \rangle \\ \langle a_1^* a_2 \rangle & \langle |a_2|^2 \rangle \end{pmatrix} \tag{2.15b}$$

where the brackets $\langle \rangle$ imply an average of the enclosed quantity over all the atoms of the ensemble. For the conditions of the experiment described above the density matrix would simply comprise a tabulation of the probabilities of finding an atom in each of its states; that is, only the diagonal elements containing the absolute magnitude squared of the various expansion coefficients (such as $\langle |a_1|^2 \rangle$) would appear. The off-diagonal elements (such as $\langle a_1 a_2^* \rangle$) involving products of different coefficients designate the extent to which the system of atoms is coherently excited and therefore capable of manifesting quantum interference effects. If the off-diagonal elements vanish, there can be no single-atom quantum interference effects. *And*, if the vapour is sufficiently rarefied

---

[29] If the two excited states have the same lifetime, then the characteristic decay rate is the inverse of that lifetime. If the lifetimes of the two states are different, then the fluorescence is describable as a sum of two exponentially decaying curves.

so the mean separation between atoms in the cell is much larger than an optical wavelength, the previously described type of quantum interference between the states of two (or more) atoms cannot occur either.

Surprisingly, the atoms can still give rise to a strange nonlocal type of quantum interference. Suppose the first experiment were in New York. Imagine an identical experiment, the 'mirror-image' of the first, set up in London; photons arrive regularly with random distribution over the same two frequencies at this station, too, and excite the same type of atoms into one or the other of its excited states. The London observer measures the fluorescent light intensity following each pulse and deduces, like the observer in New York, that the atoms are incoherently excited. However, if those two separated observers were to compare their results, they might notice something remarkable.

What links the two separated experiments is that the arriving photons are produced 'back-to-back' (i.e. with correlated linear momenta) by the same source which shall be located, let us say, midway between each station (Figure 2.8). (The location of the laboratories in London and New York is solely for the purpose of dramatising an unusual quantum

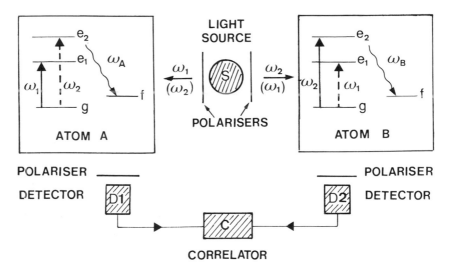

Figure 2.8. Schematic configuration of a 'long-distance' quantum beat experiment. Pairs of photons of angular frequency ω_1 and ω_2 are emitted 'back-to-back' by source S and excite widely separated absorbing atoms A and B into one or the other of two excited states e_1, e_2. The spontaneous emission from each atom separately recorded at detectors D1 and D2 manifests the simple exponential decay in time characteristic of incoherently excited systems. The joint detection probability, however, oscillates in time at the Bohr transition frequency for the two excited states.

effect. I am not concerned with problems attendant to placing a light source somewhere over the Atlantic Ocean, or with the fact that the surface of the Earth is curved!) Although the direction of emission of a photon of either frequency is random, there is one requirement that must be met: if a photon of frequency ω_1 is emitted in a certain direction, then a photon of frequency ω_2 is emitted in the opposite direction. Thus, possibly unknown to the two observers, each excitation of an atom into state e_1 in New York is accompanied by an excitation of an atom into state e_2 in London, and vice versa.

As far as each local observer is concerned, the atoms are incoherently excited into energy eigenstates from which they should decay exponentially in time. Looked at globally, however, there are again two indistinguishable quantum pathways:

$$
\begin{array}{lll}
& \text{Atom A:} & g + \text{photon } \omega_1 \to e_1 \to f + \text{photon } \omega_{A1} \\
\text{Process I} & & \\
& \text{Atom B:} & g + \text{photon } \omega_2 \to e_2 \to f + \text{photon } \omega_{B2} \\
& & \quad\quad\quad\quad\quad\quad\quad\quad\quad\quad\quad\quad\quad (2.16) \\
& \text{Atom A:} & g + \text{photon } \omega_2 \to e_2 \to f + \text{photon } \omega_{A2} \\
\text{Process II} & & \\
& \text{Atom B:} & g + \text{photon } \omega_1 \to e_1 \to f + \text{photon } \omega_{B1}
\end{array}
$$

The emitted photons are labelled by the emitting level *and* atom, since different atoms can have different velocities relative to the stationary observers. Note again that the final atomic state f does not, as before, have to be the initial state, since both atoms undergo excitation and decay.

To realise that these two processes are occurring, however, the two observers would need to compare, or correlate, the optical signals that each measures locally. Suppose the New York observer at detector D1 counts a certain number of fluorescent photons emitted within a time interval Δt_1 about the time t_1; likewise, the London observer at detector D2 counts photons in the time interval Δt_2 about time t_2. By multiplying these counts together (electronically), averaging over repeated trials and varying the detection times, the two observers can construct the joint probability $P(t_1, t_2)$ for receipt of two photons separated by the time interval $|t_1 - t_2|$. The theoretical treatment of such an experiment involves not the individual density matrices of the two separated ensembles of atoms, but rather the density matrix for the *entire* system of atoms. However far apart the two collections of atoms may be, and however independently they may go about their business after the initial excitation, they still constitute a single quantum system. The density matrix for this total system contains both diagonal elements (characterising the populations of the various states) and off-diagonal elements (signifying coherence terms that can give rise to quantum interference phenomena).

The amplitudes for the two processes in (2.16) interfere with the consequence that the jointly detected fluorescence signal, which takes the general form,

$$I(t_1,t_2) \sim A(t_1,t_2) + B(t_1,t_2)\cos\{\omega_{12}[(t_2 - t_1) - (r_B - r_A)/c]\}, \quad (2.17)$$

manifests a quantum beat at the Bohr frequency ω_{12} as a function of the delay between photon detections. Here the term $A(t_1,t_2)$ contains the matrix elements for the independent excitation and exponential decay of the two atoms to final state f; the factor $B(t_1,t_2)$ contains the matrix elements for the correlated excitation and decay of the two atoms.

The long-distance two-atom beats are insensitive to the motion of the atoms, as is seen from the fact that the atomic velocities do not appear in relation (2.17). Why is it that the quantum beats produced by the local two-atom system are strongly affected by the Doppler effect, whereas beats issuing from the correlated excitation of two distant atoms are not? In the first case (local two-atom system), the beat originates from the interference of photons potentially emissible by *different* atoms; the beat frequency is then given by the difference of two optical frequencies, each of which can independently span the full Doppler width produced by the distribution of atomic velocities. In the second case (nonlocal two-atom system), however, a beat arises from interference between the atom A amplitudes of Processes I and II in the sequence of events (2.16); corresponding interference between the atom B amplitudes also leads to this same beat frequency. Since the interferences always involve amplitudes for photon emissions by the *same* atom, the resulting quantum beat is largely unaffected by Doppler shifts.

A significant feature of the oscillatory quantum interference term in relation (2.17) is that both the time delay, $\Delta t = (t_2 - t_1)$, and the difference in retardation times, $\Delta t_r = (r_B - r_A)/c$, are multiplied *not* by the optical frequencies of the emitted photons, but by the much smaller beat frequency ω_{12}. For the quantum beat to persist when the signal is averaged over the finite surfaces of the two separated detectors, the phase $\omega_{12}\Delta t_r$ must be small compared with about 1 radian as the optical path lengths from all A atoms (in New York) to all points of detector

D1, and from all B atoms (in London) to all points of detector D2, vary. This, however, does not pose a severe experimental restriction at all. For a beat frequency ω_{12} of about 10^8 per second, for example, one can have $|r_B - r_A| < \sim 300\,\text{cm}$, which is much greater than a wavelength $(\sim 10^{-5}\,\text{cm})$ and is quite easily realised in the laboratory. As long as the distribution of retardation times satisfies the criterion

$$\omega_{12}|r_B - r_A|/c < 1 \qquad (2.18)$$

it is immaterial how far apart the two collections of atoms may be; since the two atoms of the correlated pair interact with distinctly different photons, the 'size' of the photon is inconsequential.

It will be recognised by now that the phenomenon of long-distance beats illustrates in the context of *bound* electron states correlations of a nature similar to those manifested by *free* electrons in the 'quantum interference disappearing act' described in Chapter 1. In the terminology invented by Schrödinger, the atoms of the two separated ensembles are in 'entangled' quantum states – multi-particle states that cannot be expressed as a product of the states of single particles. How atoms that undergo random excitation and exponential decay when observed locally can correlate their activity with other atoms engaged in correspondingly random behaviour an arbitrary distance away is not explicable in terms of any classical mechanism. Entangled states give rise to quantum phenomena not only unaccountable within the framework of classical physics, but that sometimes seem bizarre even by the expectations of quantum physics when those expectations are based on the study of single-particle or uncorrelated multi-particle systems.

To Schrödinger the feature of entanglement was the most characteristic property of the wave mechanics he created, but it was not a feature which pleased him. 'Measurements on separated systems', he wrote, 'cannot directly influence each other – that would be magic'[30].

---

[30] W. Moore, *Schrödinger* (Cambridge University Press, Cambridge, New York, 1989) 310.

3

And yet it moves: exotic atoms and the invariance of charge

3.1 A commotion about motion

If I had to describe in a word what physics is all about, I am tempted to say, 'motion' – as construed, of course, in a sufficiently broad sense to include not only the movement of particles, but also such phenomena as the flow of fluids, the propagation of waves, the conversion of heat to work, or the transitions between quantum states. The word 'physics', according to my unabridged English dictionary, traces its origin to a Greek term meaning 'natural things', which seems appropriate enough, as far as it goes. Actually, in some ways it goes too far, for one can accommodate all the life sciences in that definition – and, in fact, the term 'physic' once meant medicine or the art of healing. The Japanese refer to physics as 'butsurigaku', derived from 'study of transformation'. That also has some good points, but reminds me too much of chemistry. I will stay with 'motion'.

The concept of motion is fundamental in ways that go well beyond its role in physics. It underlies our very perception of reality. When Zeno of Elea propounded his paradoxes[1] in the fourth century BC, it was not merely to induce mental paralysis in his fellow mathematicians, but rather to spread the doctrine of his mentor, the mystic philosopher Parmenides, who stressed that the world of sense is nothing but illusion. Thus thought Zeno:

If motion, which pervades everything, can be shown to be self-contradictory, and hence unreal, then everything else must assume the same unreal quality. By

---

[1] Zeno proposed four famous paradoxes purportedly showing that motion cannot occur. In essence, to move from one point to another, an object must first traverse one-half the total distance. Similarly, every half interval can be decomposed into two further half intervals that must be traversed. The object, according to Zeno, cannot cover an infinite number of spatial intervals in a finite amount of time, and hence cannot move.

convincing people of the unreality of motion, I can ... successfully discredit the world of the senses[2].

Today, the fall of an apple or the movement of a planet, while probably still a mystery to the majority of laymen, does not ordinarily provoke deep philosophical discussion, much less a conceptual crisis, among physicists. Although not necessarily easy to describe mathematically, the motion of a macroscopic object is at least in principle a decidable proposition. One can tell whether it occurs or not. Look up at the Moon for a while on a clear night; its position in the sky changes. It moves round the Earth. Or does the Earth move around the Moon? In any event, something is moving, and that is a fact, Zeno notwithstanding.

Zeno was born some 2500 years too soon, for, if he enjoyed paradoxes, he would have loved quantum mechanics! Among other things, quantum mechanics disabuses us of the certainty of motion – of the perception of motion through our senses. In contrast to the observable motion of macroscopic-sized objects, one cannot 'see' an elementary particle move. For an object to be seen, it must either emit light or be illuminated. Whereas the reflection of light from the Moon or the emission of light from a firefly will scarcely alter the object's subsequent motion, the interaction of light with an elementary particle can change its behaviour drastically. One can, of course, illuminate an elementary particle with 'softer', i.e. less energetic, photons that do not substantially change its momentum. The energy of a photon, however, is inversely related to its wavelength, and the illuminated particle cannot be localised to within a region more sharply defined than the wavelength of the light used for viewing. The detailed movement of an elementary particle that has emitted or scattered soft photons would be lost in an indistinct blur. At the quantum level, motion cannot be seen directly; it must be inferred from physical theory.

Heisenberg, a seminal contributor to the creation of quantum mechanics who had given much thought to the nature of physical theory, was fairly definite about what should *not* go into one[3]:

... it seems necessary to demand that no concept enter a theory which has not been experimentally verified at least to the same degree of accuracy as the experiments to be explained by the theory.

[2] Edna E. Kramer, *The Nature and Growth of Modern Mathematics* (Hawthorn, New York, 1970) 578.
[3] Werner Heisenberg, *The Physical Principles of the Quantum Theory* (Dover, New York, 1930) 1.

Unfortunately, as Heisenberg, himself, recognised,

. . . it is quite impossible to fulfil this requirement, since the commonest ideas and words would often be excluded.

Words like orbit, trajectory, velocity – familiar terms in the Newtonian lexicon defining the classical conception of motion.

There would be no problem, Heisenberg contended, if physicists would only remain content, for example, with the images of particle tracks on photographic plates, such as those made in 1911 by C. T. R. Wilson[4], and not attempt to 'classify and synthesize' the results, or to 'establish a relation of cause and effect between them'. Without a causal description, however, how could one ever know if something was moving? Each image of an object, like the arrow in one of Zeno's paradoxes, would lie suspended and temporally disconnected from the images that came before and after.

But surely an elementary particle moves. If not, then what is one really to make of the Wilson photographs (Figure 3.1)? These photographs show the explicit tracks of alpha rays (helium nuclei) and beta rays (electrons) that passed through the supersaturated water vapour of a cloud chamber. The trails were formed from minute droplets of water that have condensed about molecules ionised by collisions with the energetic charged particles. Is not each track the trajectory of some particle that has made its way through the chamber?

Technically no – not a trajectory. The continuity of the tracks produced by numerous, but discrete, particle collisons and ionisations is illusory. Furthermore, with each random collision the subsequent motion of the particle is changed in an unpredictable way. A definite location and velocity at each instant of time cannot be assigned to the ionising particle. Nevertheless, quantum mechanical analysis does establish a picture of events at the microscopic level that conforms to expectations based on classical mechanics to within limits set by the Heisenberg uncertainty relations. From the wave function of the ionising particle the most probable value of the particle location at any instant can be inferred, and, in the absence of subsequent disturbances, this locus of points traces out the classically predicted straight line path. Each successive collision with a water molecule, however, modifies the wave function and increases the uncertainties with which the location and linear momentum of the particle can be known. The angular devi-

---

[4] C. T. R. Wilson, On a Method of Making Visible the Paths of Ionising Particles Through a Gas, *Proc. R. Soc. Lond.* **A85** (1911) 285.

Figure 3.1. Tracks of alpha particles formed in a Wilson cloud chamber. (W. Heisenberg, *The Physical Principles of the Quantum Theory* (Dover, New York, 1930) 5.)

ations of orbital segments between successive collisions depend on the relative linear momenta of the ionising particle and the atomic electrons with which it interacts. For sufficiently energetic and massive particles, like the alpha particles in Wilson's experiments, the ionised molecules will effectively lie on straight lines; the paths of the much less massive beta particles are irregularly curved.

For all practical purposes the Wilson tracks furnish an adequate record of particle motion even if one cannot assign to each point an instantaneous coordinate and velocity. No harm is done in believing that somewhere within each track a particle passed by. But what about the electrons *within* an atom; can one tell if a bound electron is moving?

From the semiclassical point of view embodied in the Bohr model of the hydrogen atom, an electron bound to a nuclear centre of charge $+Ze$ moves about the nucleus in a circular orbit of principal quantum number n with a speed

$$v_n/c = Ze^2/n\hbar c. \tag{3.1}$$

The dimensionless combination of constants $e^2/\hbar c \sim 1/137$ will be recognised as the Sommerfeld fine structure constant α_{fs} which sets the scale for the interaction of charged particles with electromagnetic fields. Although nonrelativistic ($v/c \sim 0.007$), the speed of the electron in a ground state hydrogen atom ($Z = n = 1$) is predicted by the Bohr model to be some two million metres per second, by no means a trivial quantity when gauged by ordinary experience.

In the quantum description of the hydrogen atom, however, the situation is not so simple. For one thing, corresponding to the uncertainty relation between coordinate and linear momentum is a rotational counterpart between angular momentum and angular location. Dynamical quantities such as linear momentum, angular momentum, energy, etc., which in classical mechanics can be regarded as properties of an individual particle, have a twofold significance in quantum mechanics. On the one hand, they represent mathematical expressions – or operators – that obey well-defined algebraic relations independent of any particular physical system. On the other hand, they refer to the mean values to which these operators lead when applied to the wave function of specific quantum systems. Whether the wave function and the associated dynamical variables actually characterise one particle or a large collection (an 'ensemble') of identical particles ideally prepared all in the same way is a matter of debate among physicists concerned with the foundations and interpretation of quantum theory.

In any event, an uncertainty relation can be established between any two measurable quantities for which the corresponding quantum mechanical operators do not commute. For example, the familiar Heisenberg uncertainty principle relating the uncertainties (technically the statistical variances) in coordinate and linear momentum is readily derived from the noncommutativity of the coordinates and linear momentum operators (the equation inscribed on Max Born's tombstone). Strictly speaking, there is no quantum mechanical operator for angle as there is for linear coordinate, and the establishment of a corresponding uncertainty relation between angle and angular momentum has been somewhat problematical. Nevertheless, several such relations have been proposed[5], all leading to the following consequence. Since the wave function of a hydrogenic electron – in contrast to that of a free electron – is characterised by a sharp (definite) angular momentum, the electron angular coordinate is totally delocalised about the nucleus. It

[5] P. Carruthers and M. M. Nieto, Phase and Angle Variables in Quantum Mechanics, *Rev. Mod. Phys.* **40** (1968) 411.

would seem, therefore, that the concept of an orbit or trajectory within the atom is of little, if any, significance.

A second feature of the quantum mechanical hydrogen atom contrasting starkly with the corresponding Bohr planetary model is that the angular momentum of the 1S ground state is zero. Classically, a particle with zero angular momentum either passes through the axis of rotation or has zero velocity. The quantum calculation of the mean electron orbital radius (for a state with principal quantum number n and atomic number Z) yields in accord with the Bohr model

$$r(n,Z) = n^2 a_0/Z, \qquad (3.2a)$$

where

$$a_0 = \hbar^2/me^2 \qquad (3.2b)$$

is the Bohr radius, about 5×10^{-9} cm for an electron of mass $m = 9 \times 10^{-28}$ g and charge $e = -4.8 \times 10^{-10}$ esu. Is the electron moving?

The calculation of the electron velocity is in some ways an undefined problem, for it is not velocity, but momentum, that plays a key role in the formulation of quantum mechanics. If one utilises the nonrelativistic relation between velocity and linear momentum of a particle with mass m,

$$v = p/m, \qquad (3.3a)$$

the quantum calculation leads to a null velocity. However, in contrast to classical mechanics, this does not mean that the electron *speed* is necessarily zero. Calculation of the kinetic energy

$$K = mv^2/2 = p^2/2m \qquad (3.3b)$$

of the ground state electron results in a root-mean square speed the same as that of relation (3.1). Matters become yet more confusing when one employs the relativistic electron theory of Dirac rather than the nonrelativistic Schrödinger theory. The velocity operator in the Dirac theory, defined as the time rate of change of the particle coordinate operator[6]

---

[6] In quantum mechanics the time derivative of an operator is calculated from the commutator of that operator with the Hamiltonian, H, which is ordinarily the sum of the operators representing the kinetic and potential energies. To obtain an explicit expression for the velocity operator defined in relation (3.3c) once the Hamiltonian is known, one must evaluate $v = -(i/\hbar)[xH - Hx]$.

$$v = dx/dt \qquad\qquad (3.3c)$$

has but two allowable eigenvalues: $\pm c$; this would seem to indicate that the electron can only move at the speed of light! Yet the relativistic calculation of the kinetic energy again leads (in the nonrelativistic limit) to a particle speed effectively equivalent to relation (3.1). What is to be made of all this?

Following Heisenberg's warning, one is forced to renounce all hope of visualising the motion of an electron in the energy eigenstates of an atom. Although such states – referred to as stationary states because their properties do not vary in time – may have nonvanishing expectation values of kinetic energy, angular momentum and other dynamical variables, there is still *no* sequential connection between neighbouring points in the resulting electron probability distribution.

Recognising this helps to eliminate a number of potentially paradoxical situations. Except for the 1S ground state, the radial distribution of the electron in every other hydrogenic stationary state has nodes; that is, for certain calculable distances from the nuclear centre the probability of finding the orbiting electron is exactly zero. Since the probability of finding the electron at all other points is nonvanishing, the probability distribution consists of disconnected regions. For example, the radial distribution of an electron in the 2S state consists of two disconnected regions separated by a spherical surface of radius $2a_0$ (twice the Bohr radius) on which the electron probability is zero. Given that there is but one bound electron, one might be tempted to enquire how the electron can move from the inner to the outer region if it can *never* be found at a point exactly $2a_0$ from the nucleus!

The answer is simply what has been stated before: the properties of a stationary state do not correspond to a causal description of particle motion. Since two points, however close, of a stationary state probability distribution do not represent points on a particle trajectory and provide no information at all as to how an electron may have passed from one point to the other, the presence of nodes leads to no paradoxical behaviour. According to an ensemble interpretation, we may regard the stationary states as representative of the statistical properties of a large number of similarly prepared atoms. Imagine (never mind how!) photographing the instantaneous location of the single 1S electron in one million hydrogen atoms produced by the same source. The composite radial distribution of the electrons would resemble the stationary state radial distribution predicted by quantum mechanics. Moreover, since the act of photographing the atoms perturbs the subsequent motion of the electrons, one would not be able to predict from the photograph of a particular electron where it will be an instant later.

The statistical interpretation enables us to understand as well how the 1S electron can have zero angular momentum and a nonvanishing mean speed and orbital radius. This is again what one might expect of the average angular momentum of a great number of atoms with electrons moving in randomly oriented planar orbits, clockwise or counterclockwise in equal measure. None of these properties characterises the sequential movements of an individual electron.

In marked contrast to the alpha and beta particles moving through a Wilson cloud chamber, the bound electrons in an atom leave no tracks by which to infer movement. Does a bound electron actually move? Does this even matter?

3.2 The electric charge of a moving electron

One of the attributes of particles that is in some ways both familiar and mysterious is that of electric charge. The theory of quantum electrodynamics provides a comprehensive and (as far as experiment has been able to confirm) correct description of the interaction of charged matter with electromagnetic fields. And yet, curiously enough, we do not know exactly what charge is, only what it does. Or, equally significantly, what it does not do.

Electric charge does not, on balance, change. The conservation of electric charge is one of the most strictly observed conservation laws of physics. To my knowledge no reproducibly documented violations have ever been reported. Moreover, it is *not* simply a question of global charge balance, as, for example, in a process by which an electron is created at one end of the laboratory and a positron at the other end. Charge conservation is local; there is to be no violation in any spacetime region within limits set by quantum mechanical uncertainties. Conceptually, the conservation of electric charge can be understood as arising from a special kind of symmetry in the laws of electromagnetism – the requirement that the basic equations of motion be unaffected by a phase transformation of the fundamental fields[7].

---

[7] As discussed previously in the context of the Aharonov–Bohm effect, the fundamental fields of quantum electrodynamics are the vector and scalar potentials and not the electric and magnetic fields. A function known as the Lagrangian is constructed from these basic fields. If the Lagrangian remains invariant when each field is multiplied by an arbitrary phase factor of the form $\exp(i\phi)$, where ϕ is a real infinitesimal constant, the conservation of electric charge results.

There is another, perhaps even more profound, sense in which electric charge doe not change: it is independent of its velocity. Although often taken for granted, this is a rather remarkable fact of nature, for many properties of a particle *do* depend on velocity. The apparent mass or inertia of a particle, for example, increases with particle speed; as the speed of the particle approaches that of light, an increasingly greater force is required to effect a given incremental increase in speed. To accelerate a massive particle to the speed of light would require an infinitely large force – and so cannot be done. The charge on a particle, however, does not change at all; it is said to be a Lorentz invariant.

The significance of the Lorentz transformation, named for the Dutch physicist H. A. Lorentz who discovered it in the course of developing a classical theory of the electron, was first recognised by Einstein who derived the relations in a much more fundamental way – independent of any theory of matter – through consideration of measurements of space and time. The Lorentz transformations enable two observers moving relative to one another to compare their measurements of spatial and temporal intervals and consequently to relate all other dynamical quantities (acceleration, force, energy, momentum, etc.) that are based on, or in some way connected to, space-time measurements. The importance of these relations goes well beyond the domain of mechanics. To be considered viable, a physical theory must at the least be compatible with the special theory of relativity and expressible in a form that remains invariant under a Lorentz transformation. The theory is then said to be Lorentz covariant. This means that there is no preferred reference frame – no special state of motion of an observer – for which the theory is valid. If the theory is valid, it must be recognised as such by all observers in uniform motion with respect to one another.

The space-time properties of the electromagnetic field are largely determined by the requirement of covariance *and* the Lorentz invariance of electric charge. The invariance of charge, however, unlike charge conservation, is not at present known to follow from any deeper principle, but must be taken at the outset as an experimental fact. It is conceivable, if certain types of particle decays are allowed, that charge invariance may be connected with charge conservation. For example, to maintain exact charge conservation in the disintegration of a proton into a positron plus neutral particles, the proton and positron must have equal charges irrespective of their speeds. Relativistic quantum theory requires that particles and antiparticles have charges of exactly the same magnitude. Hence, if protons decayed to positrons, then the charge of the proton and the charge of the electron would also have to be equal in magnitude. This seems to be the case although I know of no reproducible observation of proton decay.

Since charge invariance is one of the conceptual pillars of electromagnetism as we know it, the solidity of the empirical foundation upon which it rests is no trivial matter. Just how do physicists know that electric charge is independent of velocity?

One of the strongest arguments advanced for the Lorentz invariance of charge is the electrical neutrality of atoms and molecules. The essence of the argument is as follows. Suppose the charges of two elementary particles (e.g. the electron and proton) at rest are equal in magnitude. Then, were the Lorentz invariance of charge *not* valid, there would be a charge imbalance for the bound system within which these particles are in relative motion. If motion were to have an effect on the magnitude of charge, one could not expect exact cancellation of the nuclear and electronic charge in composite systems as different as the hydrogen atom, the helium atom and the hydrogen molecule (H_2).

In a stationary hydrogen atom, as discussed above, the speed of the 1S electron relative to the proton is calculated to be $v \sim 0.007c$. From relation (3.1), one would infer that the ground state (1S) electrons in a helium atom ($Z = 2$) should move twice as fast as the electron in hydrogen. Actually, of even more interest is the speed of a proton in a helium nucleus composed of two protons and two neutrons. A simple heuristic argument shows that the protons in helium are moving with relativistic speeds within the potential well that binds them by means of the strong nuclear interaction. The uncertainty in the linear momentum of a particle confined to a spatial region of size r is about h/r (where h is Planck's constant). The characteristic size r of the helium nucleus is on the order of 10^{-13} cm. Taking h/r as an estimate of the maximum value of linear momentum of a proton of mass $M = 1.67 \times 10^{-24}$ g, and equating it to the classical relativistic expression

$$p = Mv\gamma, \tag{3.4a}$$

where

$$\gamma = [1 - (v/c)^2]^{-\frac{1}{2}} \tag{3.4b}$$

leads to a proton speed of about $0.8c$, over fifty times greater than the electron speed. (If the nonrelativistic expression $p = Mv$ were used, the calculated speed of the proton would exceed the speed of light.) The electrons in a ground state hydrogen molecule are in 'molecular

orbitals' formed from the atomic 1S states and have a speed comparable
to that of the electron in a hydrogen atom. The maximum speed of the
two protons, oscillating about their equilibrium separation of nearly 1
ångström (10^{-8} cm) at a frequency on the order of 10^{13} Hz, is approximately 6×10^5 cm/s or $2 \times 10^{-5} c$ – considerably less than that of the
two protons in the helium nucleus.

In a set of experiments performed over thirty years ago, J. G. King[8]
established that the fractional difference in charge between electrons
and protons in helium atoms and in hydrogen molecules is zero to within
a few parts in 10^{20} or one hundred billion billion! The experimental
technique is as ingenious as it is simple. In brief, the gas whose charge
was to be measured was allowed to escape from an electrically insulated
metal container attached to an electrometer; a 'de-ioniser' swept out of
the gas stream any ions or free electrons. If there were a charge
imbalance in the remaining gas molecules, then the outflow of gas would
result in a current flow to the container that would be registered by the
electrometer. To within an experimental uncertainty leading to the
above awe-inspiring limit on electron–proton charge equality, no such
current was found. It would seem that the Lorentz invariance of charge
rests on indisputably firm ground.

But does it really? Although the demonstrations of atomic and molecular neutrality are convincing, there is still one conceptually untidy step
in the chain of reasoning to charge invariance. How do we know that the
experiments actually examine the variation in charge with particle
speed? How do we know, that is, that within the atom (or within the
nucleus) the particles are truly moving? Call it a matter of professional
ethics, if you will, but can one in good faith claim that an electron in a
stationary state is moving when he wants to justify the Lorentz
invariance of charge, and then forgo a space-time description of this
motion in order to avoid quantum mechanical paradoxes? I wonder
what Heisenberg would say.

While the visualisation of motion is by no means necessary or even
relevant to the consistency of quantum mechanics, it does play a certain
important role in special relativity. In the latter, one must be able to
imagine the placement of clocks and metre sticks in different inertial
reference frames for the purpose of performing space-time measurements. If the Lorentz invariance of charge is to be inferred from the
state of motion of bound elementary particles, then – although the

[8] J. G. King, Search for A Small Charge Carried by Molecules, *Phys. Rev. Lett.* **5** (1960)
    562.

equation of motion must necessarily be quantum mechanical – it must still be meaningful to conceive of a Lorentz transformation relating the rest frame of a particle to the rest frame of another particle or of a stationary observer. However, given that a bound electron cannot even in principle be located without alteration of its state of motion, is such a conception meaningful?

The question of electron motion within an atom first started gnawing at me when I was an undergraduate student a few years after the King experiments were performed. I wondered whether there was any phenomenon at all exhibited by a bound electron that could manifest directly some element of the kinematics of special relativity. It seemed to me that, despite the fact one could not picture the motion of a bound elementary particle, somehow the question ought to be answerable in the affirmative. The assertion that a particle is in motion relative to a stationary observer has an observable physical consequence: a clock moving *with* the particle must exhibit the relativistic effect of time dilation (also called time dilatation). In the words of the mnemonic familiar to those who study special relativity: 'Moving clocks run slow'. A time interval determined from a moving clock would appear *shorter* than the same interval measured with a system of clocks relatively at rest[9].

If the electron were a classical charged particle orbiting a centre of force, this effect would in principle be manifested in the Doppler shift of radiation emitted from different points in the electron trajectory. As in a binary-star system, there would occur both a blue and a red shift for any angle of observation other than at 90° to the plane of orbital motion. The electron, of course, is not a classical particle – and the immediate proof of that is the very existence of atoms. If the electron radiated as described, it would spiral into the nucleus, and (as may be demonstrated from the Larmor formula (2.7a)) the atom would collapse in about 10^{-11} seconds! In the quantum mechanical atom radiation is not continuous, but occurs only when the electron undergoes a transition between states; the radiation frequency, unrelated in general to the calculable orbital frequency, is not Doppler shifted unless the entire atom is moving.

There is, however, an alternative and peculiarly quantum mechanical

---

[9] Imagine two stationary clocks A and B placed on the ground a distance d apart. A third clock C moving at a speed v parallel to the ground passes over A at which moment the A and C times are recorded. When C passes over B, the B and C times are likewise instantaneously recorded. Although the difference in A and B readings shows the passage of a time interval d/v, the C clock will have advanced by only $d/v\gamma$.

clock associated with an elementary particle, namely its natural lifetime. But the electron, as far as one knows, is a stable particle; there is no other negatively charged particle of lower mass to which it can decay. The electron lifetime should be infinite, and no effect of motion on it would be observable. But suppose an electron *could* decay. Would its lifetime be lengthened if it were bound in an atom?

3.3 The exotic atom

At first acquaintance a so-called exotic atom may seem like a chimera of atomic physics – but it is, in fact, quite real. Since an inevitable comparison with a planetary system is made whenever atoms are discussed, imagine again looking at the Moon one night and seeing it vanish in a burst of light. Something similar can happen in an exotic atom – except that the burst might well be one of electrons and neutrinos (ghostly spin-$\frac{1}{2}$ particles without charge or mass that interact extremely weakly with all matter). Or imagine the Solar System with one of the planets orbiting *inside* the Sun. That, too, can happen in an exotic atom. An exotic atom is formed when one of the electrons of an ordinary atom is replaced by another more massive (and unstable) negatively charged elementary particle.

The story is oft repeated that I. I. Rabi, when first apprised of the existence of the muon in the 1930s, grumbled, 'Who ordered *that*?', in displeasure at the increasing complexity of nature. The number of 'elementary' particles known today is so large that most can scarcely be considered elementary, but the muon still is. With a mass of about 207 times the electron mass, the muon is described by particle physicists as essentially a heavy electron[10]. There is one critical difference, however: the muon has a finite lifetime. In its rest frame the muon lasts about 2.2 microseconds before transforming via a weak interaction (like the β-decay of a nucleus) into an electron, an electron antineutrino and a muon neutrino. Like the electron, the negative muon can bind to a positive atomic nucleus to form an atom – one of the exotic atoms. (To the positive electron, or positron, there also corresponds a positive muon.)

A few microseconds may seem like a miniscule amount of time for a particle, and the atom it forms, to stay around. In comparison to initial

---

[10] I recall a seminar many years ago with the title 'Is the Moon a Heavy Electron?' The misprint on the flyer must surely have attracted people who came to learn of the marvellous new astronomical 'discovery'.

expectations, however, the muon lifetime was anomalously long. In 1935, based on analogy with the transmission of the electromagnetic force by photons, Hideki Yukawa predicted the existence of a carrier of the strong nuclear force. The electromagnetic force is known to be of infinite range as a consequence of the zero rest mass of the photon. Knowing that the strong force is of short range (about the size of a nucleon, 10^{-13} cm), Yukawa was able to predict that the mass of the sought-for nuclear particle should be about two hundred times the electron mass. Not long afterward, a particle of that approximate mass was observed in the cosmic ray showers that reached the Earth's surface. But *this* particle did not interact strongly with atomic nuclei; if it had, its existence would have been a fleeting 10^{-23} second, and it would never have passed through the Earth's atmosphere to be detected at ground level. The discovery was actually the muon[11].

According to relations (3.2a,b), the ground state orbital radius of a negative muon bound to a positive nucleus of charge $+Ze$ should be smaller than that of the corresponding electron orbit by the ratio of the electron and muon masses

$$r_\mu(Z) = (m/m_\mu)a_0/Z \sim a_0/(207Z). \tag{3.5}$$

The orbital speed, however, depends only on particle charge and should be the same, $v/c \sim Z\alpha_{fs}$, for both the muon and electron. Hence, a bound muon with lifetime t_0 can complete about $vt_0/(2\pi r_\mu) \sim 3 \times 10^{12}Z^2$ revolutions before undergoing weak decay. In terrestrial terms, this is the equivalent of over a thousand billion years, many times longer than the current age of the Earth. The muonic atom would therefore seem to be a reasonably stable system with well-defined ground state.

The existence of exotic atoms was inferred by Enrico Fermi and Edward Teller in 1947 from the different behaviour of positive and negative muons coming to rest in iron or graphite[12]. Positive muons, repelled by the positive atomic nuclei, were expected to decay naturally at their characteristic rate producing 'disintegration electrons'; negative muons – at a time when the muon and the predicted Yukawa particle (pion) were not yet recognised as distinctly different particles – were expected to be captured by the atomic nuclei and not give rise to

[11] Yukawa's particle, termed the pi-meson or pion, was found in 1947. There are three varieties. The neutral pion, π^0, decays naturally with a lifetime of about 10^{-16} second; the charged pions, π^+ and π^-, have a natural lifetime of about 10^{-8} second.

[12] E. Fermi and E. Teller, The Capture of Negative Mesotrons in Matter, *Phys. Rev.* **72** (1947) 339. A 'mesotron' is the old designation of 'meson' – or medium-mass particle. At the time of the article the distinction between muons and pions was not yet clear.

electrons. This expectation was fulfilled for iron. In graphite, however, disintegration electrons emerged equally abundantly from both positive and negative muon decays in sharp disagreement with contemporary expectations. To explain this anomaly, Fermi and Teller assumed that the negative muon was captured into a Bohr orbit from which it, too, decayed naturally rather than by nuclear capture. They showed that a negative muon initially captured into a high-lying atomic state cascaded down into the ground level – generally referred to as the K shell – in a time interval on the order of 10^{-12} second for atoms in condensed matter and 10^{-9} second for atoms in a gas – i.e. in an interval short compared with the muon lifetime. Fewer than about 10^{-2} muons decay from states other than the ground state.

Once the muon reaches the K shell, it is in an orbit some 200 times smaller than that of a K shell electron; largely unaffected by the surrounding shells of atomic electrons, the muon can be treated to good approximation as if it were the only bound particle. The radius of a nucleus of atomic number Z and mass number A (number of protons and neutrons) is usually represented by the formula

$$R = A^{\frac{1}{3}} r_0, \tag{3.6}$$

where $r_0 = 1.3 \times 10^{-13}$ cm is about one-half the so-called classical electron radius, e^2/mc^2. From relations (3.5) and (3.6) it is seen that for muonic silver ($Z = 47$, $A = 108$), the muon 1S orbital radius of about 5×10^{-13} cm lies at the periphery of the nuclear surface. In muonic uranium ($Z = 92$, $A = 238$) the ground state muon orbit of radius about 3×10^{-13} cm lies well within the uranium nucleus (radius $\sim 8 \times 10^{-13}$ cm).

When captured by a nucleus, the negative muon interacts with a proton to give rise to a neutron and muon neutrino. The probability of this process increases with nuclear charge because the degree of overlap of the muon and nuclear wave functions is correspondingly greater as the orbital radius is smaller. For $Z = 11$ (muonic sodium) the rates of weak decay and nuclear capture are approximately equal; for heavy nuclei the mean lifetime of a bound muon is essentially determined by the nuclear capture process. Nevertheless, since the end products of the two processes are different, one can experimentally distinguish them and study the natural decay of bound muons by monitoring the rate of production of decay electrons.

The natural lifetime of the muon constitutes an ideal clock for testing the time dilation effect of special relativity. Indeed, such a test was first made over half a century ago in a now classic experiment by Bruno Rossi and David Hall[13] whereby the decay rate of cosmic ray muons was shown to depend on particle speed in accordance with the relativistic relation

$$t_\mu = \gamma(\beta)t_0. \tag{3.7}$$

Here t_μ is the observed lifetime (reciprocal of the decay rate) of muons moving relative to a stationary observer with speed parameter $\beta = v/c$, t_0 is the lifetime in the muon rest frame, and $\gamma(\beta)$ is the dilation factor given by relation (3.4b).

I recall learning of the time dilation of muons in an educational film[14] depicting a modified repetition of the Rossi–Hall experiment some twenty years later at about the time the question of electron motion first occurred to me. In the film, the lifetime of energetic muons ($v/c\sim0.995$) measured at the top of Mount Washington in New Hampshire and at sea level in Cambridge, Massachusetts, was found to be lengthened by nearly a factor of nine. I wondered: would the lifetime of a muon in the K shell of a muonic atom also be lengthened?

A quick estimate of the dilation factor of the bound muon lifetime can be made simply by substituting the Bohr speed $v/c = Z\alpha_{fs}$ into relation (3.4b). This would suggest, for example, that in muonic uranium, where $v/c\sim0.41$, the decay rate of a K shell muon should be only about 0.77 that of a free muon at rest. This is not, however, a rigorous way to deduce the properties of a quantum system which should in principle be determined by means of the expectation value of an appropriate operator. But what operator would correspond to time dilation? Since the relativistic expression for the total (mass+kinetic) energy of a free particle of mass M is

$$K = Mc^2\gamma, \tag{3.8a}$$

and since the Hamiltonian operator H that governs the time-evolution of a quantum system is the sum of K and the potential energy V, it seemed to me reasonable to define a relativistic time-dilation operator by

$$\gamma = (H - V)/Mc^2. \tag{3.8b}$$

[13] B. Rossi and D. H. Hall, Variation of the Rate of Decay of Mesotrons with Momentum, *Phys. Rev.* **59** (1941) 223. We still have 'mesotrons', rather than muons, since at the time of the experiment the pion was not yet discovered.
[14] D. H. Frisch and J. H. Smith, Measurement of the Relativistic Time Dilation Using μ-Mesons, *Am. J. Phys.* **31** (1963) 342.

The expectation value of γ for the $1S_{\frac{1}{2}}$ ground state of a point Coulomb potential $V = Ze^2/r$, evaluated by means of the exact relativistic Dirac wave functions, led to the satisfying result

$$\gamma = [1 - (Z\alpha_{fs})^2]^{-12} \qquad (3.8c)$$

in agreement with the semiclassical relativistic theory of the Bohr atom. For intermediate and heavy muonic atoms in which the muon orbit lies closer to or within the nuclear surface, the potential seen by the muon can be decidedly different from that of a Coulomb potential. In fact, a muon orbiting deep inside a nucleus with uniform distribution of positive charge will experience a potential similar to that of a harmonic oscillator. Nevertheless, once the potential V is known, relation (3.8b) can be used to determine the relativistic dilation of a bound state lifetime.

Coincidentally, although I was not to realise this until some years afterwards, accurate determinations of the lifetime of a number of moderate and heavy muonic atoms[15] had been made shortly after King's experimental test of electron and proton charge equality. In experiments performed with the 200 MeV synchrocyclotron at the University of Liverpool, negative muons travelling at a speed of about $0.94c$ were brought to rest in various targets, and the resulting time distribution of decay electrons was monitored. This distribution is governed by the familiar exponential decay law (there are no quantum beats here!)

$$N(t) = N_0 \exp(-t/t_\mu), \qquad (3.9)$$

where $N(t)/N_0$ is the fraction of undecayed muons at time t. Before each selected target was used, an experimental run was made with a carbon target which furnished an effective measurement of the muon decay rate under circumstances comparable to free muon decay (since $Z = 6$ leads to $\gamma \sim 1.00$).

So ... do bound muons move? Summarised in Table 3.1 below is the experimentally observed ratio $R = t_0/t_\mu$ of the bound and free muon decay rates. Also shown is the theoretically expected ratio $R = 1/\gamma(Z)$ where, for ease of calculation, the point Coulomb potential was assumed. In the case of heavy muonic atoms, where the time dilation effect is greatest, a more appropriate model of the electrostatic potential should in principle be employed. Nevertheless, it is clear that the overall trend shows basic agreement between experiment and theory. Let us

---

[15] I. M. Blair *et al.*, The Effect of Atomic Binding on the Decay Rate of Negative Muons, *Proc. Phys. Soc.* **82** (1962) 938.

Table 3.1. *Experimental and theoretical values of R(Z)*

Element	Z	$R(Z)_{theo.}$	$R(Z)_{expt.}$
V	23	0.99	1.00±0.04
Fe	26	0.98	1.00±0.04
Ni	28	0.98	0.96±0.04
Zn	30	0.98	0.93±0.04
Sn	50	0.93	0.87±0.04
W	74	0.84	0.78±0.04
Pb	82	0.80	0.86±0.04

not reach any hasty conclusions, however. Actually, to be completely above board it must be noted that the muon decay rate is influenced not only by relativistic time dilation, but also by dynamical and statistical effects incurred through the binding. The two principal effects, which modify the decay rate in opposite ways, are termed the electron Coulomb field effect and the phase space effect. In the first, the positively charged nucleus attracts the electron emitted in the muon decay and thereby produces a greater overlap of the muon and electron wave functions near the nucleus than would otherwise be the case for a freely propagating electron (described by a plane wave). The electron Coulomb field effect thereby enhances the muon decay rate. From the perspective of classical physics it may seem strange that the end product (an electron) formed *after* the decay has occurred can have an antecedent effect on the rate of occurrence of that very process. However, the weak decay of a particle with concomitant production of new particles is not a classically explicable process. In the phase space effect, the volume of phase space – in essence the number of quantum mechanical states – accessible to the decay products of a bound muon is restricted by the electrostatic binding force in comparison to that of a free muon. Hence, the phase space effect reduces the decay rate.

Detailed calculations[16] of the decay rate that take account of all contributing processes suggest that at least in the light- and medium-mass elements there is a fortuitous cancellation between the Coulomb field and phase space effects; the observed difference in lifetime between bound and free muons could then be principally attributed to the kinematic effects of special relativity. For the heavy elements as well, the accord between theory and measurement would be impaired

[16] R. W. Huff, Decay Rate of Bound Muons, *Annals of Physics* **16** (1961) 288.

were the effect of time dilation not included. It does seem that bound muons are in some kind of motion after all.

As for the electron – well, it is just a light muon, and my doubts have been dispelled. Who can even begin to imagine what strange scenes an electron encounters in its endless unfathomable trek around, near and possibly through the nucleus? Trajectory and velocity has it none; and yet it moves.

3.4 Epilogue: marking time with planetary atoms

Though the exact movement of an electron in an atomic stationary state cannot be pictured, recent developments in the investigation of highly excited atoms nevertheless suggest an enticing possibility: the creation of a bound electron wave packet following (in a probabilistic sense) a classical Keplerian orbit about the nucleus[17]. The properties of such a state differ vastly from those of a completely delocalised stationary state for which only the mean orbital radius may correspond to the radius of the corresponding Keplerian (i.e. Bohr) orbit.

Rydberg levels, it should be recalled, are closely separated with the energy interval between any two adjacent electronic manifolds varying inversely as the third power of the principal quantum number. Under appropriate circumstances, it is possible to excite an electron (e.g. with a broadband laser) into a linear superposition of Rydberg states comprising a broad distribution of principal quantum numbers. The resulting wave packet is then localised with respect to its radial coordinate.

For a superposition of Rydberg states with *low* angular momentum, for which the corresponding classical elliptical trajectory is highly eccentric, the electron probability distribution is delocalised with respect to the angular coordinates. When close to the inner turning point of its orbit, where its acceleration is greatest, the electron can emit light most strongly. Far from the nucleus, the electron behaves more like a free particle and cannot emit radiation. One can therefore study the motion of the delocalised electron wave packet by monitoring the time dependence of the emitted light. The periodic return of the electron to the ion core – until the wave packet ultimately decays – would be signalled by bursts of light emission at time intervals given by the classical orbital period.

There is much interest in producing an electron wave packet with a

[17] G. Alber and P. Zoller, Laser Excitation of Electronic Wave Packets in Rydberg Atoms, *Physics Reports* **199** (1991) 231.

broad linear superposition of Rydberg states of *high* values of angular momentum. Such a wave packet would be localised with respect to both radial and angular coordinates and should orbit the nucleus very much like a classically bound particle. The classical trajectory would be nearly circular, however, and at first glance it would seem that the rate of light emission should no longer depend on location of the electron in its orbit. How would the periodic motion be discerned experimentally?

Although the electron in a hydrogen atom experiences a purely $1/r$ Coulomb potential, the electrostatic potential field through which an alkali atom Rydberg electron moves includes a small $1/r^4$ contribution resulting from core polarisation and relativistic effects. This deviation from the Coulomb potential causes the wave packet to *precess*. That is, after repeated revolutions of the electron, the orientation of the packet relative to some fixed direction in the laboratory, slowly changes. This effect is quite similar to the precession of the orbit of Mercury about the Sun as a result of an analogous deviation from the $1/r$ potential of Newtonian gravity predicted by Einstein's theory of general relativity.

An electron wave packet comprising many states of large principal and orbital quantum numbers, but with only the highest magnetic quantum numbers contributing, will have a well-defined alignment relative to some arbitrary, but fixed, axis. Suppose the Rydberg atoms were subjected to a pulsed electric field which can ionise the highly excited electron, i.e. provide enough energy to separate it entirely from the atom. The rate of ionisation would then depend on the orientation of the wave packet relative to the electric field, and be greatest when the packet is aligned along the field. In principle, therefore, the precession of the wave packet could be inferred from the time dependence of the ionisation signal. Because of the very long time scale of the precession (order of milliseconds), the effect has not yet been observed. Its accomplishment, however, is literally only a matter of time.

4

Reflections on light

4.1 Exorcising a Maxwell demon

On the outside wall of the University, facing the Rue Vauquelin, is a small plaque with the words:

EN 1898 DANS UN LABORATOIRE
DE CETTE ECOLE
PIERRE ET MARIE CURIE
ASSISTES DE GUSTAVE BEMONT
ONT DECOUVERT LE RADIUM

The institution is the Ecole Supérieure de Physique et Chimie Industrielles (E.S.P.C.I.) in Paris. The historical laboratory in which the Curies discovered radium – a draughty wooden shed dismantled long ago – stood in the courtyard practically below the window of my office in what is today the Laboratory of Physical Optics, a building of vintage appearance itself. A small engraved stone marker at the edge of a parking place is all that indicates the location of the Curies' shed. Nevertheless, whenever I looked out my window, I could imagine watching Marie and Pierre tirelessly processing the tons of pitchblende ore that arrived periodically from the St Joachimsthal mines in Bohemia.

Except for singular occasions like the E.S.P.C.I. centenary anniversary in 1982, the thought of the Curies, if it exists at all, is probably a distant one in the minds of the researchers and students scurrying up steps and through corridors of the many buildings that border the courtyard. But history can be fascinating and instructive. Appointed to the Joliot Chair of Physics at the E.S.P.C.I., I was curious to know more about the distinguished son-in-law Frédéric Joliot, in whose eponymous suite of rooms furnished by the University I was living with my wife and young children. Did he and his wife, Irène Curie, both of whom served as assistants to Marie, return home covered with radioactive dust? Were

Irène's cookbooks radioactive like those of her mother? Was my family now breathing in these toxic exhalations of a past age of scientific glory? Happily this was not the case. To my relief, Joliot was an 'ancien élève' and not a professor at the E.S.P.C.I., and he and Irène never lived in the 'Apartement Joliot'. There can be solace in history – at least if one keeps the facts straight!

It was not the discovery of radium, nor for that matter anything connected with nuclear physics, that brought me to the E.S.P.C.I., but another event much less well known and celebrated. There, in the 1960s, my French colleague, Professor Jacques Badoz and his students developed the photoelastic modulator (PEM), an ingenious and versatile optical device for examining the property of light known as polarisation[1]. The years may have passed, but, in a tradition seemingly typical of the French 'grandes écoles', members of the original group were still there, no longer students, of course, but researchers leading their own groups. A centre of expertise for constructing and using PEMs, the E.S.P.C.I. Laboratory of Physical Optics was as likely a place as any to exorcise what Jacques and I had come to regard as our 'Maxwell demon'.

The Scottish physicist, James Clerk Maxwell, is one of the giants of nineteenth-century classical physics. His theory of electromagnetism unified under one set of laws the hitherto separate branches of physics constituting electricity, magnetism and optics. Maxwell's great synthesis provided the foundation upon which contemporary quantum physicists continue to build in their efforts to unify all known physical interactions. A major contributor as well to the science of thermal phenomena (thermodynamics and statistical mechanics), Maxwell once described how, by means of a molecular-sized sentient being (the Maxwell demon), the Second Law of thermodynamics might be circumvented. The Second Law of thermodynamics, of which there are several different but equivalent versions, states in essence that a spontaneously occurring process increases the disorder of the universe. In point of fact, such a violation of physical law is not possible – but all that is actually irrelevant to this essay. The issues discussed here have nothing to do with thermodynamics; they concern instead a devilishly difficult experiment with light, and the implications of the experiment for Maxwell's electromagnetic theory[2].

---

[1] M. Billardon and J. Badoz, Modulateur de Biréfringence, *Comptes Rendus Acad. Sci.* **262** (1966) 1672.

[2] At the end of one particularly frustrating afternoon, when the light source was

That light behaves like a wave had been demonstrated by means of diffraction and interference experiments long before the culmination of Maxwell's work in the 1860s. However, the early developers of wave theories of light did not know what was actually 'waving'. The pinnacle of Maxwell's achievement in this area was to deduce from his four basic laws of electricity and magnetism the equation of a wave whose calculable speed of propagation, whether in vacuum or in a material medium, was numerically equal to the corresponding speed of light. This speed was expressed in terms of properties of the medium – the electric permittivity (or dielectric constant) ε and the magnetic permeability μ – that can be determined by electric and magnetic experiments not in any way directly involving the properties of light. From Maxwell's theory (applied to an optically isotropic, homogeneous medium) it readily followed that the index of refraction n, which is the ratio of the speed of light in vacuum to the speed of light in the medium, should be equal to $\sqrt{(\varepsilon\mu)}$.

Light, then, was shown to be an *electromagnetic* wave: the propagation – even through materially empty space – of synchronously oscillating electric and magnetic fields. In the absence of matter, or in the presence of an optically isotropic material, the oscillating electric and magnetic fields of a light wave are perpendicular to each other and to the direction in which the wave is propagating.

That there could still have been any doubt that Maxwell's theory provided a complete and correct description of classical electromagnetic phenomena came as a surprise to me when I first became aware of it in the early 1980s. Nevertheless, as my colleague Jacques remarked in referring to another surprising revelation, 'Il y a donc encore des taches blanches sur les cartes du continent scientifique . . .' In the case at hand, however, these blank spaces were found, not in some far-flung impenetrable valley of the scientific continent as one might have expected (e.g. the realm of quantum gravity), but rather in the developed urban area of physical optics. At issue were fundamental principles governing the reflection of light – a subject that one might well have thought was laid to rest over a century and a half ago.

A good controversy in science is ordinarily a noisy affair, at least

fluctuating, an amplifier failed, the monochromater somehow vanished into a student laboratory, and background electrical noise was comparable to shouting at a soccer match, Jacques turned to me and asked rhetorically, 'Quel démon nous fait faire cette expérience?' The answer came to both of us simultaneously: Maxwell's. (An operational device that seemingly mimics a true Maxwell demon will be discussed in Chapter 5.)

within the discipline affected, accompanied by academic teeth-gnashing and *ad hominem* aspersions. (The well-informed reader has but to recall the recent controversies over the extinction of the dinosaurs and the alleged discovery of 'cold fusion'.) But that was not the case here; the controversy passed quietly through the journal pages without creating any furore at all.

And yet the issues involved were fundamentally of momentous import. Within the context of more-or-less routine investigations there arose questions with extraordinary implications. Had one found an area of *classical* optics that fell outside the scope of Maxwell's theory? Did Maxwell's theory lead to violation of the law of energy conservation? Were the Maxwellian boundary conditions – the mathematical expressions describing the behaviour of electromagnetic waves at an interface between different media – wrong? Quiet and unnoticed, the controversy effectively embraced the restructuring of classical electromagnetism.

What theoretical inadequacy or experimental observation could possibly have led to such far-reaching implications? The problem, in fact, took its origin in theoretical attempts to answer a deceptively simple question: how does light reflect from a left- or right-handed material? The statement of the problem may draw a sceptical raise of the eyebrow from a reader unfamiliar with chemical structure who recalls childhood jests about left-handed spanners. But such chiral, or handed, materials exist, however, and manifest intriguing phenomena collectively known as optical activity.

My interest in light reflection from chiral media arose as the natural extension of a series of experiments begun several years earlier addressing a completely different controversial issue centred on what outwardly would appear to be an even more provocative question: can a *greater* amount of light be reflected from a surface than is incident upon it? Although the correctness of Maxwell's theory of electrodynamics was not formally at issue, the experiments nevertheless rigorously tested that theory in a domain far removed from common experience in which prior theoretical and experimental efforts gave rise to confusing and conflicting results. Although I did not articulate the solution as such at the time, in a way it was an exorcism of another Maxwell demon.

My studies of the interaction of light at the surface of different media did not lead to new or modified laws of electrodynamics. But I had learned once again that what seemed to be well known was not necessarily well understood, and even so venerable a subject as classical optics still had its surprises.

4.2 Enhanced reflection: how light gets brighter when it is up against a wall

A light beam incident upon a transparent material is partially transmitted through the surface and refracted (i.e. deviated from its original direction), and partially reflected from the surface. The exact division of light energy between these two processes was first worked out in the early 1820s – that is, long before the electromagnetic theory of light – by the French physicist and engineer, Augustin Fresnel, whose name, like that of Maxwell's, is associated with a variety of inventions, discoveries and principles. Along with the Englishman Thomas Young, Fresnel was a major contributor to the perception of light as a wave-like phenomenon.

Fresnel regarded light waves as a type of elastic wave like that of sound passing through air or of ripples spreading on the surface of water. Since all elastic waves require a medium, Fresnel assumed that the light propagated through an extremely tenuous hypothetical medium, the aether, that permeated all space and penetrated all objects. Not until many years later, after Einstein developed the theory of special relativity in 1905, were most physicists fully prepared to dispense with the concept of the aether. Nevertheless, Fresnel's elastic theory enabled him to predict or account for many aspects of the behaviour of light. In the course of his investigations of light polarisation, Fresnel deduced the amplitudes (relative to an incident light wave of unit amplitude) of light specularly reflected and refracted at the surface of a transparent medium. (Specular reflection is 'mirror-like' reflection from a smooth surface, in contrast to diffuse reflection from a rough surface.) These amplitudes are ordinarily designated the Fresnel relations or Fresnel coefficients. The relative intensity of the reflected or transmitted light is proportional to the square of the corresponding Fresnel coefficient and is termed the reflectance or transmittance, respectively. Ironically, although Fresnel pioneered the investigation of light polarisation, it was precisely this attribute of light that his elastic theory was incapable of treating adequately.

Within the framework of electromagnetic theory, the polarisation of light is defined by the orientation of the oscillating electric and magnetic fields. Light waves, for which the electric and magnetic fields oscillate in planes, are said to be linearly polarised; the direction of polarisation is by convention the direction of oscillation of the electric field. Light waves, for which the electric and magnetic fields rotate about the direction of propagation, are said to be circularly (or, more generally, elliptically) polarised; the sense of the rotation, clockwise or counterclockwise to someone looking towards the light source, defines the type of circular polarisation as right or left, respectively.

Light waves propagating through a vacuum (or any optically isotropic medium) are transverse waves: the electric field, magnetic field and propagation direction are all mutually orthogonal. It was one of Fresnel's signal achievements to recognise the transverse nature of light, but it was this very point that constituted a serious flaw in his derivation of the Fresnel relations. These equations are correct, but they would *not* have been had Fresnel implemented consistently the boundary conditions that pertain to the passage of an elastic wave from one medium to another. Correctly applied, the elastic theory of light gives rise to a refracted wave that is not purely transverse, but has a longitudinal component, i.e. corresponds to an oscillation of the medium parallel to the direction of wave propagation. There is no experimental evidence for the existence of such a longitudinal wave.

It was Maxwell's theory of electromagnetism that provided the first self-consistent derivation of the Fresnel relations (without the extraneous longitudinal wave)[3], and the experimental verification of these relations correspondingly constitutes an important confirmation of the *particular dynamics* of Maxwell's theory. Certain kinematical aspects of reflection and refraction, such as the laws governing the angles at which light is reflected and refracted, were well known at least since the seventeenth century, and are characteristic of any wave theory. The above emphasis on dynamics underscores the fact that the mathematical form of the Fresnel amplitudes is not characteristic of all wave theories, but depends sensitively on the specific laws governing the electric and magnetic fields.

Although reflection and refraction amplitudes can be derived for any state of light polarisation, what is ordinarily designated *the* Fresnel relations refers to two basic types of linear polarisation, s and p, defined with respect to the 'plane of incidence', i.e. the plane formed by the incident light ray (the direction along which the incident wave propagates) and a line normal (perpendicular) to the reflecting surface[4]. This plane is always perpendicular to the reflecting surface (Figure 4.1). For

---

[3] This derivation was first given by the Dutch physicist, H. A. Lorentz, in 1875.

[4] The letters s and p derive from the German words for perpendicular ('senkrecht') and parallel. The two polarisations are also referred to as TE (transverse electric) and TM (transverse magnetic).

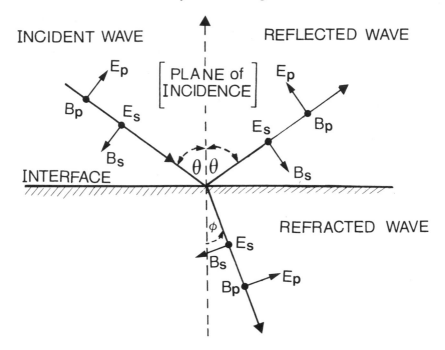

INCIDENT WAVE   REFLECTED WAVE

Figure 4.1. Geometry of waves specularly reflected and refracted at a planar interface (perpendicular to the page) between two dielectric media. The plane of incidence (lying in the page) is defined by the incident light ray and the normal to the interface. Waves are designated s-polarised or p-polarised depending upon whether the electric field is perpendicular (E_s) or parallel (E_p) to the plane of incidence. The angles of incidence and reflection (θ) are equal; the angle of refraction (ϕ) is given by Snel's law.

s-polarised waves, the electric field is perpendicular to the plane of incidence, and therefore parallel to the reflecting surface irrespective of the angle of incidence at which the light ray intersects the reflecting medium. For p-polarised waves, the electric field oscillates within the plane of incidence; the angle that it makes with the surface depends on the incident angle of the light which is ordinarily measured with respect to the surface normal.

The exact variation of the Fresnel coefficients with angle of incidence depends on the relative index of refraction of the two media forming the interface at which the light is reflected and refracted. Nevertheless, for most familiar dielectric (nonconducting) materials the following general behaviour is observed. The intensity of reflected s-polarised light at normal incidence is low and increases nonuniformly, but continuously, with the angle of incidence until it reaches 100% at exact grazing incidence (i.e. for a light ray skimming the surface at 90° to the normal direction). The intensity of reflected p-polarised light equals that of s-polarised light at normal incidence, but with increasing angle of incidence it drops to exactly 0% at a special angle called the Brewster angle[5], after which, as in the case of s-polarisation, the reflectance continuously grows to 100% at grazing incidence (Figure 4.2). An incoherent mixture in equal measure of s- and p-polarised beams generates a beam of unpolarised light.

It is very likely that nearly everyone – whether familiar with the Fresnel relations or not – has experienced in one way or another the optical phenomena embraced by these expressions. The dim image one

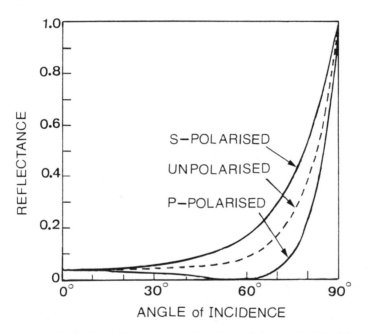

Figure 4.2. Variation in reflectance as a function of the angle of incidence for light reflected from a medium with refractive index higher than that of the medium within which the light originates. The reflected intensity of s-polarised waves increases smoothly between normal and grazing incidence, whereas the intensity of p-polarised waves drops to zero at the Brewster angle. If the refractive index of the reflecting medium is lower than that of the incident medium, then the reflectance reaches 100% at the critical angle θ_c and remains unchanged over the range from θ_c to 90°.

---

[5] The Brewster or polarising angle θ_B for light originating in a medium with refractive index n_1 and reflecting from the surface of a medium with index n_2 is determined from the relation: $\tan \theta_B = n_2/n_1$.

sees while standing before a clear window pane at night readily cor-
roborates that only a small fraction of normally incident light is reflected[6]. By contrast, the bright glare on a glossy magazine held at an
oblique angle to a light source irritatingly confirms that a large fraction
of light near grazing incidence is reflected.

Polarised light of a well-defined frequency and propagation direction
is one of nature's 'absolutes'; it cannot be reduced to anything simpler.
The light that we ordinarily encounter, such as direct sunlight or light
from incandescent and fluorescent bulbs, is largely unpolarised; the net
electric field of the constituent waves is distributed randomly in time
over all possible orientations perpendicular to the propagation direction. It is of interest to note, then, that specular reflection of *un*polarised
light at the Brewster angle results in a 100% s-polarised reflected beam!
At first glance this phenomenon may seem extraordinary – the generation of complete order out of disorder in apparent violation of the
Second Law of thermodynamics. There is no violation, however, for the
light and reflecting surface constitute an open system not in thermodynamic equilibrium (i.e. characterised by different temperatures).
Nevertheless, reflection at the polarising angle can lead to some remarkable – and easily observable – optical effects.

I can recall one example in my own kitchen where a crumpled cellophane bread wrapper lying on a smooth table and illuminated from
behind by an open window gave rise to an impressive array of colours on
the table surface. This phenomenon is an example of the 'interference
colours' produced by a birefringent material, i.e. an optically
anisotropic material for which the speed of light – and correspondingly
the index of refraction – depends on the direction of light propagation.
In the usual classroom demonstration of this effect, a birefringent
material is placed between two linear polarisers (sheets of polaroid
plastic, for example). The first polariser constrains the electric field of
the transmitted light wave to a well-defined plane. Upon entering the
birefringent material, the light wave is split into two components for
which the electric field is either parallel to, or perpendicular to, a special
symmetry axis of the material designated the optic axis. The two com-

---

[6] At normal incidence the magnitude of the Fresnel amplitude for both s- and p-polarised
light is

$$r = (n - 1)/(n + 1),$$

where n is the ratio of the refractive index of the second medium, e.g. glass, to that of
the incident medium, e.g. air. For glass in air, n is about 1.5, and the reflectance is
$r^2 \sim 0.04$.

ponents travel through the material at different speeds and thereby incur a relative phase shift by the time they emerge and recombine at the far end of the material. Because the two components are orthogonally polarised – that is, their electric fields are mutually perpendicular – they cannot interfere and manifest any effect of the relative phase shift[7]. The second polariser, however, projects the electric fields of both components onto a common transmission axis whereupon the two waves linearly superpose and interfere. Depending on the phase shift, the two components may reinforce one another and appear brighter, or interfere destructively and be eliminated. Since the phase shift varies with the thickness of the material and the wavelength (colour) of the light, the composite effect on an incident white light beam of the constructive and destructive interference of waves passing through an unevenly thick birefringent material is to produce a bright patchwork quilt of colours when the material is viewed through the second polariser.

In the 'kitchen experiment' above, the crumpled cellophane constituted a birefringent material of randomly varying thickness, but where were the two polarisers, for none had been explicitly employed? The first polariser was that of the atmosphere itself. The sunlight streaming through the open window had been polarised by incoherent molecular scattering (Rayleigh scattering) by the oxygen and nitrogen molecules in the air. This effect is greatest for skylight coming from directly overhead when the Sun itself is near the horizon; the light scattered at 90° to the incident light can in principle (though, because of depolarising mechanisms, rarely in practice) be polarised 100% with the electric field normal to the plane determined by the incident and scattered light rays. The second polariser was the smooth table surface, itself, which reflected the light (passing through the cellophane) at the Brewster angle.

There is another angle – known as the critical angle – that plays an important role in light reflection when the reflecting material has an index of refraction *lower* than that of the medium within which the light originates. When light passes, for example, from glass (with an index of refraction $n_1 = 1.5$) to water (with an index of refraction $n_2 = 1.3$) the

---

[7] The linear superposition of two light waves, represented by their electric field vectors E_1 and E_2, produces a net wave of amplitude $E = E_1 + E_2$. The interference term in the resulting intensity

$$I = |E|^2 = |E_1|^2 + |E_2|^2 + 2E_1 \cdot E_2$$

vanishes as a consequence of the scalar product if the two fields are orthogonal.

refracted rays in the water bend *away* from the axis normal to the glass–water interface in accordance with Snel's law of refraction

$$n_1 \sin\theta = n_2 \sin\phi, \tag{4.1}$$

where θ is the incident angle and ϕ is the refracted angle (both angles measured with respect to a line perpendicular to the surface). Starting with a light beam directed normally at the interface ($\theta = 0$), for which the transmitted light beam is undeviated ($\phi = 0$), and gradually increasing the angle of incidence, one eventually reaches the critical angle θ_c at which the refracted rays are parallel to the surface ($\phi = 90°$). It readily follows from relation (4.1) that the critical angle characterising a particular interface is determined from

$$\sin\theta_c = n_2/n_1 \qquad \text{(where } n_1 > n_2\text{)}. \tag{4.2}$$

For incident angles greater than θ_c relation (4.1) leads to no real transmission angle ϕ. Experimentally, no light propagates through the second medium; *all* incident light is reflected irrespective of polarisation.

The effect of 'total reflection' is readily demonstrated by a transparent container (preferably one with flat rather than rounded sides) filled with water. Looking up through the bottom of the container an observer would have no difficulty seeing his finger held just above the surface of the water. When viewed upward through one of the sides, however, the surface is no longer transparent, but appears opaque and reflective like a mirror. The effect can be particularly pronounced to someone submerged in the transparent water of a swimming pool gazing obliquely upward towards the surface and seeing, not the sky or ceiling, but objects at the bottom of the pool.

A fundamental attribute expected of the Fresnel amplitudes is that they must be compatible with the conservation of energy. The formal expression of this for a transparent medium is that the reflectance and transmittance must sum to 100% at any angle of incidence and for any polarisation. No light energy can be lost or created by reflection.

It may seem paradoxical, but it is nevertheless the case, that although all incident light energy goes into the reflected wave under conditions of total reflection, the incident wave still penetrates the surface of the reflecting medium. The transmitted wave is not a travelling wave – i.e. it does not propagate undiminished through the medium, but dies off exponentially with depth of penetration. This 'evanescent' wave transports no net energy into the medium. Nevertheless, its effects can be dramatic.

* * * * * * * * * * * *

If I were to single out one of the achievements of theoretical physics that has had the greatest impact on twentieth-century science and technology, it would be Einstein's prediction in 1917 of *stimulated* light emission[8]. Before Einstein's work, physicists and chemists were familiar – even if they did not understand the mechanism – with the production of light by *spontaneous* emission, the process that gives rise to the spectral lines of excited atoms and the fluorescence and phosphorescence of molecules. Einstein reasoned, however, that in addition to spontaneous emission there had to exist another light-creating process if a piece of matter were to reach thermodynamic equilibrium with its environment. Whereas spontaneous emission, like the name implies, ordinarily occurs without external provocation at a rate characteristic of the quantum states of the emitter[9], stimulated light emission is driven by the presence of light.

Some thirty-five years passed before Einstein's prediction was verified in the operation of the ammonia maser, a device that produced microwave radiation by stimulated emission. The term 'maser' is an acronym deriving from *m*icrowave *a*mplification by *s*timulated *e*mission of *r*adiation. Shortly afterwards, in 1960, the stimulated emission of visible (red) light was first produced in the ruby laser (substitute 'light' for 'microwave'). Today, there is hardly a branch of science, technology, industry or medicine that does not employ lasers in one capacity or another. Clearly, the need to amplify light is of great importance.

In the operation of a laser, energy is pumped – for example by optical, electrical or chemical means – into the atoms or molecules of the lasing material. The material is said to be excited, for its elementary constituents have been driven into their excited quantum states. When the number of atoms (or molecules) in an excited state is greater than that of a lower energy state to which quantum transitions are possible, the system is said to have a population inversion. Under appropriate conditions – in particular, if the wavelength falls within the emission spectrum of the atoms or molecules – a light wave propagating through the excited

---

[8] A. Einstein, Zur Quanten Theorie der Strahlung, *Physik. Zeit.* **18** (1917) 121.
[9] From the perspective of quantum electrodynamics, spontaneous emission is a type of stimulated emission induced by fluctuating electromagnetic fields of the vacuum. Confining an excited atom or molecule to a sufficiently small enclosure significantly modifies the fluctuations of the vacuum, and therefore also the spontaneous decay rate of a quantum state; see D. Kleppner, Inhibited Spontaneous Emission, *Phys. Rev. Lett.* **47** (1981) 233.

material can stimulate the release of this stored energy. Whereas the original mode of excitation (such as an electrical discharge through the lasing medium) may be, in a manner of speaking, disordered, the process of stimulated emission is a highly ordered one. A photon present in the wave stimulates an atom or molecule to emit a second photon with identical properties. In this way, upon multiple passages of the light *through* the material, stimulated emission can in principle turn a weak initial wave into an intense, collimated, monochromatic, polarised light beam.

I have emphasised the word 'through' above to accentuate a most unusual feature of a mode of light amplification proposed by Soviet researchers in 1972 that differs markedly from the laser[10]. The researchers purported to demonstrate theoretically that a light beam can be amplified by specular *reflection* from the surface of an excited substance under conditions of total reflection – that is, where *none* of the incident light propagates through the energetic medium!

The claim stimulated as much contention as light. For one thing, the mathematical analysis of this novel effect suffered a serious ambiguity. For another, experimental tests yielded degrees of amplification far different from those predicted. And thirdly, the physical mechanism for how such an amplification could occur outside the amplifying medium was not entirely clear.

Let us consider first the theory. To someone not familiar with the application of mathematics to physics, it may seem surprising that a properly conducted analysis can lead to ambiguous results. The popular (and not undeserved) image of physics as a mathematically rigorous science would seem to imply that, given the equations of motion for some system, one could in principle always (although not necessarily easily) solve them – and, if the equations are correct, then the solutions will accurately describe the system. No two ways about it! Unfortunately, the situation is rarely that simple. The equations that govern a physical system – and which are ordinarily differential equations relating the temporal and spatial rate of change of dynamical quantities – usually give rise to more than one solution, perhaps to an infinite number, distinguished by the choice of the initial conditions (specifying the state of the system at some fixed time) or the boundary conditions (specifying the state of the system at some fixed place).

[10] G. N. Romanov and S. S. Shakhidzhanov, Amplification of Electromagnetic Field in Total Internal Reflection from a Region of Inverted Population, *Sov. J. Exp. Theor. Phys.* **16** (1972) 209.

The problem with the foregoing study of light reflection, based on the Fresnel relations for a uniformly excited medium, was that the pertinent equations gave rise to two fundamentally different solutions. To see how this came about, let us examine the standard Fresnel relation for the reflection of s-polarised light at the interface of two ordinary *unexcited* dielectric media with respective indices of refraction n_1 and n_2 (where the light originates in the medium characterised by n_1)

$$r_s = \frac{n_1 \cos\theta - n_2 \cos\phi}{n_1 \cos\theta + n_2 \cos\phi}. \tag{4.3}$$

For any incident angle θ chosen by the experimenter, the angle ϕ at which the light is refracted in the second medium is governed by Snel's law, relation (4.1). In fact, it is not just the angle ϕ that appears in the amplitude r_s, but rather $n_2 \cos\phi$, and this may be written explicitly in terms of the incident angle as follows[11]

$$n_2 \cos\phi = [(n_2)^2 - (n_1 \sin\theta)^2]^{\frac{1}{2}}. \tag{4.4}$$

Since the index of refraction of a transparent medium is the ratio of the speed of light in vacuum to the speed in the medium, it must consequently be a positive real number. Moreover, if $n_2 > n_1$, as in the case of ordinary reflection (e.g. light originating in air and reflecting from glass), the left-hand side of equation (4.4) must also be a positive real number, and equation (4.4) leads to a unique angle of refraction ϕ for any θ within the allowed range of $0°$ to $90°$.

In the case where the refractive index of the second medium is *lower* than that of the first ($n_2 < n_1$), the right-hand side of equation (4.4) is the square root of a negative number, and therefore a pure imaginary number, for incident angles beyond the critical angle θ_c given by relation (4.2). There are two possible square roots differing by a sign, and for neither can ϕ be interpreted as an angle of refraction. Indeed, as described in the previous section, the light is totally reflected. One of the roots, which is the appropriate one for this physical system, leads to the evanescent wave. The other root, however, gives rise to a wave that grows exponentially with penetration of the medium, and, as the presently considered medium has no latent source of energy with which to augment the wave, this root must be discarded as unphysical.

If the second medium is *not* transparent – if, for example, it absorbs

---

[11] Substitute the expression, $\sin\phi = (n_1/n_2)\sin\theta$, obtained from Snel's law (equation (4.1)) into the trigonometric identity $\cos^2\phi = 1 - \sin^2\phi$ and take the square root.

light – then the situation becomes somewhat more complicated, for now the refractive index itself is a complex number expressible in the form

$$\tilde{n} = n + i\varkappa. \tag{4.5a}$$

The real part n, the ordinary refractive index, characterises the phase shift incurred by a wave as a result of its spatial displacement through the medium; the imaginary part $\varkappa$, the absorption coefficient, characterises the diminution in amplitude of the wave as a result of absorption. Thus, the amplitude of a plane wave of angular frequency ω that has propagated a distance d through an absorbing dielectric medium is proportional to

$$\exp(i\tilde{n}\omega d/c) = \exp(in\omega d/c)\exp(-\varkappa\omega d/c). \tag{4.5b}$$

If the medium is *amplifying* rather than absorbing, then the amplitude of a travelling wave grows as it propagates. As inferred from relation (4.5b), an amplifying medium is one for which the absorption coefficient $\varkappa$ is negative. In any event, when the refractive index of the second medium is complex, then equation (4.4) leads to two complex square roots (one the negative of the other) and therefore to two different expressions for the Fresnel amplitude r_s. But in this case, it is not so obvious which root to retain.

Upon substitution of the complex expression for $n_2 \cos\phi$ which has both real and imaginary parts

$$n_2 \cos\phi = q' + iq'', \tag{4.6a}$$

the resulting reflectance can be written in the form

$$R_s = |r_s|^2 = \frac{(n_1 \cos\theta - q')^2 + (q'')^2}{(n_1 \cos\theta + q')^2 + (q'')^2}. \tag{4.6b}$$

One of the two roots of equation (4.4) generates a positive q' and a negative q'' in which case it is clear from equation (4.6b) that the Fresnel coefficient is less than unity for all angles of incidence, except at grazing incidence ($\theta = 90°$) where the light is totally reflected. There is no amplification. The second root, however, leads to a negative q' and positive q'', from which it correspondingly follows that, exclusive of $90°$, the reflectance (4.6b) is greater than unity for all angles of incidence. Thus, if the second root is the correct one, light amplification is predicted at all angles, except for total reflection at grazing incidence. Is the light amplified or not? How is one to know which root to select?

In their analysis of enhanced reflection, the Soviet authors made the following arbitrary selection: choose the first root (no amplification) for angles of incidence below critical angle and the second root (amplification) for angles of incidence above critical angle. At critical angle, itself, there is a discontinuity in the Fresnel coefficient. A similar problem and resolution applies to the reflection of p-polarised light.

An examination of the actual wave forms (similar to that of relation (4.5b)) corresponding to the two roots leads to the following interpretation. The wave for which q' is positive and q'' is negative represents an *amplified* wave propagating *away* from the interface into the second (the excited) medium. This is the sort of wave one would expect to propagate within a medium with a population inversion, as in the example of the laser described earlier. The wave for which q' is negative and q'' is positive represents a *decaying* wave that originated infinitely deep within the excited medium and is propagating *towards* the interface. According to the common perception that light reflection occurs at the outside surface of the second medium, the existence of such a wave, especially under conditions of total reflection, is problematical. From where and how did this wave arise? I will return to this question later.

When determined by the foregoing arbitrary selection of roots, the Fresnel coefficients for a uniformly excited medium lead to a maximum intensity enhancement of about e^2, or less than 10. Does such an amplification of reflected light actually occur?

At about the same time the theory was published, a different group of Soviet scientists reported a most interesting experiment[12]. Light from a neodymium laser was directed at normal incidence through a glass prism in contact with a solution of organic dye (known as Rhodamine 6G). The composition of the solution was adjusted so that its index of refraction was a little less than that of the overlying glass in order that total reflection could occur at the glass–dye interface. Upon absorbing the light, the dye molecules became excited, and a population inversion was established. The excited dye molecules, undergoing spontaneous radiative transitions back to the ground level, isotropically emitted light (fluorescence). A portion of the fluorescence emerging from one end of the prism was directed by a mirror back onto the glass–dye interface at a narrow range of angles spanning the critical angle. These rays then reflected from the interface and were received at a distant photographic

---

[12] B. Ya. Kogan *et al.*, Superluminescence and Generation of Stimulated Radiation under Internal-Reflection Conditions, *Sov. J. Exp. Theor. Phys. Lett.* **16** (1972) 100; S. A. Lebedev *et al.*, Value of the Gain for Light Internally Reflected from a Medium with Inverted Population, *Opt. Spectros.* **35** (1973) 565.

film. From measurement of the extent of exposure of the film (by a densitometer) as a function of the angle of incidence, the researchers concluded that the reflected fluorescence was enhanced by about a factor of 25. In a subsequent experiment a maximum enhancement in excess of 1000 was reported!

By the time I first learned of the controversy over enhanced reflection, the waters surrounding it were rather muddied. Theoretical attempts to account for what the Soviet group may have actually observed were not satisfactory. Nor did there seem to be experiments by other researchers. Assuming the phenomenon of enhanced reflection actually existed, theorists did not in the main agree over some of its basic attributes. According to some, amplification could be produced at any angle of incidence; according to others, the phenomenon occurred only for incident angles beyond critical angle.

As the controversy bore on conceptually subtle issues of both theoretical and practical importance, one of my graduate students and I decided to examine the problem systematically. We were not interested in trying to explain an experiment performed under conditions that we only vaguely understood, and that may not have corresponded at all to the theoretical analyses worked out by others. Rather, we set out to devise an experiment where all conditions and parameters were clearly ascertainable and to compare our observations with a theoretical analysis directly applicable to that experiment.

Recognising that an infinitely deep uniformly excited medium, such as initially treated, is an idealisation that one does not encounter in an actual experiment, we examined the type of 'gain profile' – i.e. the spatial distribution of excited molecules – one is likely to engender by pumping a dye solution with a laser. If the intensity of the laser pump is not too great, so that the effective ground state lifetime (the reciprocal of the pumping rate) is much longer than the lifetime of the excited states, then the gain profile has the same spatial dependence as that of the pump beam in the dye solution. (Recall from the discussion of quantum beats that this is the condition required to avoid multiple cycles of absorption and stimulated emission during the passage of the pump light pulse.) It has long been known (and referred to as Beer's law) that the intensity of a weak light beam passing through an absorbing medium diminishes exponentially with depth of penetration z as follows:

$$I(z) = I(0) \exp(-z/\delta). \tag{4.7a}$$

The characteristic depth of penetration δ depends on the number of absorbing molecules and the effectiveness (or so-called absorption cross-section) with which a molecule can absorb a photon from the pump beam. Where molecules have been excited, the dielectric constant ε (and correspondingly the refractive index) incurs a negative imaginary part which, in consequence of equation (4.7a), also follows an exponential decay law within the medium

$$\varepsilon(z) = \varepsilon(0)[1 - i\gamma \exp(-z/\delta)]. \tag{4.7b}$$

Here $\varepsilon(0)$ is the dielectric constant of the (transparent) medium in absence of excitation and γ, the gain parameter, is a measure of the strength of the pump beam or, equivalently, the extent of population inversion.

The solutions of Maxwell's equations for the waves of s and p polarisation that propagate through a nonmagnetic ($\mu = 1$) medium with dielectric constant (4.7b) were (with no pun intended) quite illuminating. For one thing, the results tended to vindicate the Soviet theorists' arbitrary selection of roots in the case of a uniformly excited medium. Far from the interface – i.e. for a penetration large in comparison to the characteristic depth δ – the excited medium is effectively transparent rather than amplifying (since the amplitude of the pump beam is too low to effect much of an excitation). In this region a wave that had entered the surface at an angle below θ_c behaved, as expected, essentially like a plane wave travelling undiminished away from the interface. For angles of incidence above critical angle, however, the character of the wave changed; it became an evanescent wave attenuated exponentially with distance of penetration. So far, so good.

It was in the excited layer within a few penetration lengths δ of the interface that the physics became interesting. Examination of the exact solutions (which are complicated mathematical expressions involving, or resembling, Bessel functions) showed the waves to be decomposable into a linear superposition of two basic components. One component, analogous to the amplitude derived from the first root of equation (4.4), represented an amplified wave that travels away from the interface into the excited medium; the other component – as the reader may have surmised – characterised a decaying wave travelling towards the interface, such as derived from the second root of equation (4.4).

In the present case, however, no question arises concerning which component to keep, nor is there any discontinuity in the reflectance at critical angle. For any light polarisation there is only one physically acceptable solution – namely, the solution that remains finite at the interface; this solution can contain both components. The two components do not necessarily contribute to the total wave form equally, but depend on the angle of incidence. Nevertheless, their relative magnitudes are automatically provided by the theory. In order for enhanced reflection to occur, the component travelling *towards* the interface has to be present in the excited layer. This occurs for angles beyond a certain threshold angle which ordinarily lie close to the critical angle. In this way, therefore, the prediction of enhanced reflection from a uniformly excited medium can be considered justified.

Calculations of the enhancement expected for various experimental conditions believed achievable in the laboratory led to peak values of less than 10 over a very narrow range of angles – a few hundredths to a few tenths of a degree – in the vicinity of critical angle; outside this range the amplification fell off rapidly. Also, other things being equal, the enhancement increased as the critical angle approached grazing incidence. The reason for this will become clearer shortly. Was such amplification observed?

Indeed it was, although the stringent requirements posed by theory on the stability of the experiment made for no mean task. As in the Soviet experiment, we constructed a dye cell with an organic dye (Rhodamine B) as the medium to be excited (Figure 4.3). A thick fused quartz window with flat sides overlay the dye. Varying the composition of the dye solvents permitted a coarse adjustment of the critical angle to a value of about 88°; fine adjustment was subsequently achieved by careful control of the dye cell temperature. To create a population inversion, the dye was excited by light from a pulsed dye laser directed at normal incidence through the transparent quartz window, and the highly collimated, monochromatic beam furnished by a separate helium–neon laser served as a probe with which to measure the amplified reflection. After reflection from the quartz–dye interface, the probe beam was sent through a monochromater (an instrument employing a diffraction grating to remove extraneous fluorescence) and then into a photodetector.

The dye cell was mounted on a rotatable stage so that the reflectance could be measured over a range of incident angles. Since amplification was predicted for only an extremely narrow range of incident angles close to critical angle, precise control and measurement of the angle of the probe beam were critical to the success of the experiment. This was achieved by employing a second helium–neon laser located at a screen

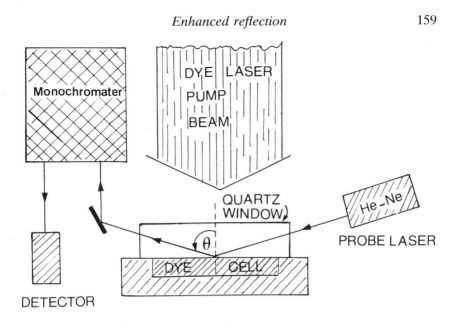

Figure 4.3. Schematic diagram of the enhanced reflection experiment. Light from a pulsed dye laser creates a population inversion in the dye solution. The probe beam from a helium–neon laser reflects at the interface between the dye solution and a quartz window (of slightly higher refractive index), passes through a monochromater which filters out fluorescent light from the dye, and is received at a photodetector. Amplification is expected for incident angles in the vicinity of the critical angle.

about five metres from the dye cell. Light from this calibration laser reflected from the quartz surface leaving a sharp red spot on the screen from which the direction of normal incidence could be inferred. By removing the photodetector and determining the location and size (~1 mm) of the probe beam spot on the screen, one could measure the angle of incidence to within one-hundredth of a degree – a precision adequate for testing the theory of enhanced reflection.

The intensity of the reflected probe beam was then measured, first with the dye unexcited and then while the dye was being pumped, and the reflectance was determined as a function of incident angle for different choices of critical angle θ_c (in the vicinity of about 88°) and penetration length δ (ranging from about 40 to 100 wavelengths of the 633-nanometre helium–neon red light). With peak amplifications that reached 200–300%, depending on experimental conditions, we were greatly satisfied to find that the results agreed well with the theoretical expectations of our model (Figure 4.4a,b).

(a)

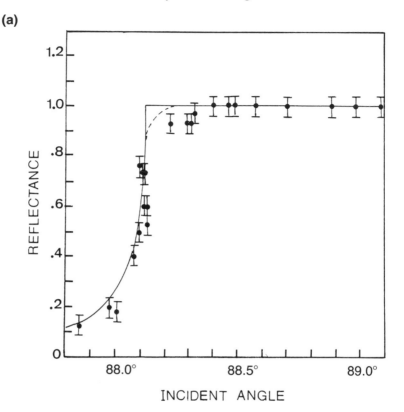

INCIDENT ANGLE

Figure 4.4. (a) Reflectance as a function of incident angle for light reflection from the unexcited dye (of Figure 4.3) at a wavelength for which the dye is largely transparent. The critical angle is at 88.12°. As expected, there is no amplification. The solid line denotes the theoretical reflectance of a transparent medium. The broken line shows the small modification that occurs when the finite divergence of the probe beam is taken into account. (b) Example of the reflection curve obtained from a dye solution with population inversion. With a critical angle of 88.79° and characteristic penetration length of about 58 wavelengths (of 633 nanometre helium–neon laser light) the intensity of the reflected beam is amplified nearly twofold. The solid line shows the theoretically predicted reflectance.

Although the amplification of light by reflection is now well established, the phenomenon may yet raise some puzzling questions in the minds of those who have not thought about the matter of light reflection before. How does amplification actually occur if, in contrast to the laser, light does not propagate through the excited medium? As I pointed out before, total reflection does not imply that the incident light

(b)

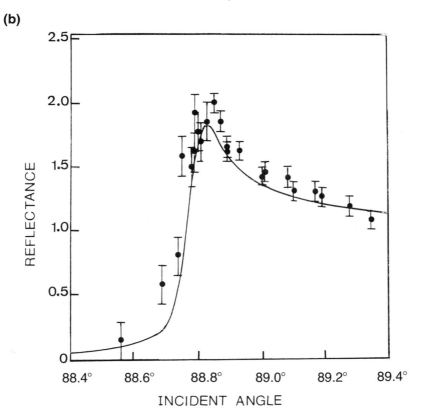

is in no contact whatever with the excited molecules. There is the evanescent wave whose depth of penetration, depending on the angle of incidence, can be substantial. Heuristically speaking, enhanced reflection may be attributable to stimulated emission by this evanescent wave.

Considering again the simplest case of reflection from a uniformly excited medium, I note that the component of the plane wave travelling through the medium in a direction *normal* to the surface has the simple form

$$E(z) \sim \exp[\mathrm{i}(\omega/c)(n_2 \cos\phi)z]. \tag{4.8a}$$

Making use of equations (4.2) and (4.4), one will see that for angles of incidence beyond critical angle (where $n_2 \cos\phi$ is a complex number) the above expression reduces to that of an exponentially damped wave

$$E(z) \sim \exp[-z/\delta'], \tag{4.8b}$$

where the penetration depth δ' for a wave of wavelength $\lambda = 2\pi c/\omega$ is

$$\delta' = \frac{(\lambda/2\pi n_1)}{[(\sin\theta)^2 - (\sin\theta_c)^2]^{\frac{1}{2}}}.$$ (4.8c)

As the angle of incidence θ approaches critical angle θ_c, the penetration δ' becomes infinitely large. The degree of enhancement depends on the extent to which the incident wave – even though totally reflected – penetrates the medium. Hence, greater amplification is expected for incident angles close to critical angle. Note, too, that the incident wave also has a component travelling *parallel* to the surface. The closer the critical angle is to grazing incidence, the more of the excited medium the parallel component traverses – just like sunlight passes through more of the Earth's atmosphere when the Sun is near the horizon than near the zenith.

Clearly the above 'explanation' is not complete, for one may well wonder why there is no enhancement for angles below critical angle where a transmitted travelling wave could conceivably stimulate the medium to radiate. The enhancement occurs only when there is present inside the excited medium that wave component travelling towards the interface. From where does it come?

The equations describing the process of enhanced reflection from a laser pumped medium are complicated, and it must be admitted frankly that not every feature predictable by the theory is correspondingly amenable to a simple interpretation. Nevertheless, there is a subtle and significant issue here that bears profoundly on the nature of light reflection in general which, when once appreciated, may help clarify the existence in the reflecting medium of a seemingly puzzling wave form.

Why, for example, does the reflection of light at the outside surface of a material depend at all on the optical properties of the interior region? The answer, by no means trivial to demonstrate, is given by what is termed the Ewald–Oseen extinction theorem. The basic idea is that at an atomic level the reflected and transmitted waves, although apparently generated at the boundary, actually originate from *within* the reflecting medium by the coherent radiation of molecular dipoles that have absorbed (extinguished) the penetrating incident light. The waves radiated by individual molecules superpose constructively in the directions for which reflected and transmitted waves are predicted to exist by macroscopic physical optics. For other directions the waves superpose destructively. Thus the reflected wave, although existing in medium 1, bears the imprint of the optical properties of medium 2. And in a similar way, the refracted wave, whether travelling or evanescent, originates ultimately within the interior of the medium and not exclusively at the interface.

To my knowledge, this microscopic picture of light reflection has been implemented rigorously only in a few tractable cases such as reflection at the interface of transparent media. Even then the analysis is not simple[13]. Were a corresponding microscopic treatment of enhanced reflection to be given, I think it very likely that a satisfactory molecular explanation of the wave form within the gain region would emerge. In any event, it is worth emphasising that the decomposition of a given wave form into various components is a mathematical stratagem that can usually be effected in different ways for different purposes. In reality, there is only *one* wave of specified frequency and polarisation within the amplifying region and no problematic wave propagating towards the surface from the infinite depth of the material.

* * * * * * * * * * *

Having satisfied myself that the amplification of *linearly* polarised light by reflection from an excited medium was possible and followed in a self-consistent way from the laws of classical electrodynamics, I wondered next whether one could selectively amplify *circularly* polarised light by the same method. There are certain materials that interact asymmetrically with left and right circularly polarised light when unexcited, and I expected that they would do the same when pumped to higher quantum states. I also expected that the theoretical analysis of this process would be easily effected – at least for the special case of a uniformly excited medium. All I had to do was start with the appropriate Fresnel relations for a transparent unexcited material and then, following the approach taken by the Soviet theorists, replace the real refractive indices with complex ones with negative imaginary parts.

I was wrong. Six years would pass before I could return to the problem of enhanced reflection from a medium that reflects left- and right-handed light differently. I was to discover first that the simplest case I imagined – a problem that ought to have been solved during the last century – seemed to have no physically acceptable solution!

4.3 Left- and right-handed reflection

That light waves could exist in right- and left-handed forms was a bold hypothesis initially proposed and experimentally confirmed around 1825

---

[13] See, for example, M. Born and E. Wolf, *Principles of Optics*, 4th Ed. (Pergamon, Oxford, 1970) 104–8.

by Fresnel as part of his investigation of a curious phenomenon today referred to as optical rotation. Many naturally occurring substances, corn syrup for example, have the capacity to rotate the plane of vibration of a transmitted linearly polarised light beam. Some fourteen years earlier the French physicist, Dominique F. J. Arago, first observed this effect – the nature of which he did not understand – in the passage of linearly polarised sunlight along a particular axis (called the optic axis) of a quartz crystal. The sunlight had been polarised by reflection from glass at the Brewster angle. By viewing the light through a plate of Iceland spar (calcite), Arago saw two solar images in complementary colours. He had seen such colours before when the light passed through plates of mica or gypsum instead of quartz, but in those cases the colours of the images changed when the plate was rotated in its plane. The orientation of the quartz plate about the optic axis, however, had no effect on the images.

It was another French 'optician', Jean Baptiste Biot, who recognised shortly afterwards that the effect resulted from the rotation of the linear polarisation of the light in quartz. Biot carried out extensive investigations of the phenomenon discovered by Arago – his first written memoir in 1812 read before the Institut de France covered some 400 pages! – showing that optical rotation occurred not only in crystals, but also in liquids such as turpentine and oils of laurel and lemon, and in their vapours. On occasion the demonstrations were accompanied by spectacular optical effects of an unanticipated nature, as when Biot set up his gas-phase optical rotation experiment in an ancient church then serving as the orangery for the house of peers. Turpentine vapour, issuing from a boiler, was conducted into a thirty-metre long iron tube with glass ends. Just when the effect of optical rotation was beginning to be observable, the boiler exploded setting fire to the church! Unfortunately, Biot was not able to measure the extent of the optical rotation (although I have no doubt that city officials readily quantified the extent of the damage).

In general, the degree to which the plane of polarisation is rotated is proportional to the quantity of substance through which the light travels and inversely proportional to the wavelength of the light. For most natural products the rotation is usually modest, perhaps a few degrees or tens of degrees per millimetre of substance, although there are materials for which the rotation can be much larger[14]. (In a special class

---

[14] A comprehensive treatment of optical activity, including history, experimental techniques, and tables of data, is given in the classic work by T. M. Lowry, *Optical Rotary*

of materials known as cholesteric liquid crystals, the rotations can be enormous, on the order of $100\,000°$ per millimetre.) What is the explanation of optical rotation?

The essence of Fresnel's interpretation, still valid today, is that, upon entering the material, linearly polarised light is decomposed into two coherent beams of opposite *circular* polarisation. The right- and left-handed waves propagate with different speeds and incur a relative phase shift. Emerging from the far side of the material, the two circular polarisations – no longer rotating in synchrony – superpose again to yield a linearly polarised beam with rotated plane of polarisation (Figure 4.5). For a light beam of wavelength λ passing through a substance of thickness d, the degree of rotation θ can be expressed as

$$\theta = \pi(n_L - n_R)d/\lambda, \tag{4.9}$$

where n_L and n_R are the different indices of refraction for left and right circular polarisations.

One of the most striking demonstrations of optical activity that I know – and which would no doubt have greatly pleased Fresnel – con-

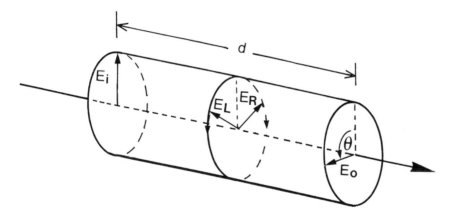

Figure 4.5. Optical rotation of a plane-polarised wave in an optically active medium. The electric field (E_i) of the incident linearly polarised wave is a superposition of left (E_L) and right (E_R) circularly polarised components which advance through the medium at a rate depending on the respective refractive indices n_L and n_R. Upon leaving the medium, the superposition of the phase-shifted fields E_L and E_R results in a linearly polarised field E_0 rotated by angle θ with respect to E_i. The degree of rotation is proportional to the circular birefringence ($n_L - n_R$) and the sample length (d).

Power (Longmans, Green, and Co., London, 1935), reprinted in 1964 by Dover Publications (New York).

sists of passing the red beam of a helium–neon laser through a long vertical transparent glass tube of corn syrup, the bottom of which (where the light enters) is covered by a rotatable linear polariser. Without the linear polariser the tube of syrup is more or less uniformly reddish along its length when regarded from the side. With the polariser, however, the light appears distributed around the axis of the tube much like the red spiral of a barber pole. Indeed, by rotating the polariser in one sense or the other, one can make stripes of light wind upward or downward around the tube axis.

This remarkable effect is a combination of both optical rotation and molecular light scattering. At the plane of entry through the linear polariser, the electric field of the light is oscillating along a well-defined axis perpendicular to the light beam. As the beam propagates upward through the syrup, this axis is progressively rotated. From a classical perspective, the electric field of the light wave, which is oscillating at some 10^{14} cycles per second, causes electrons of the sugar molecules in the syrup to vibrate about their equilibrium positions and therefore to radiate electromagnetic waves of the same frequency as that of the incident light. By this process of absorption and re-radiation, the sugar molecules redirect or scatter a part of the incident light. The theory of molecular light scattering – the same theory (Rayleigh scattering) that accounts for the polarisation of skylight – predicts that the light scattering is greatest at right angles to the axis along which the electric charges are oscillating. Thus, as the syrup continuously turns the electric field of the advancing light wave, it turns, as well, the orientation of the induced electric dipoles in the medium and the direction in which the incident light is maximally scattered. Hence the observed 'barber pole' effect.

Optical rotation is one of a complex of phenomena, more generally termed optical activity, the mechanisms of which entail an asymmetric interaction with right- and left-handed light. For a substance to interact asymmetrically with the two forms of circularly polarised light, it must itself be built of units that have a handedness or chirality. (A chiral object, like a glove, is one that in general cannot be superposed on its mirror image.) This can occur in two basically different ways.

A material may be optically active because its fundamental chemical unit, or molecule, is chiral. This is the case for the corn syrup which is composed of sugar molecules with a characteristic right or left handedness. In most organic compounds there is at least one carbon atom whose chemical bonds are directed outward towards the vertices of a tetrahedron, i.e. with an angle of about 109° between any two bonds. If a molecule contains one or more 'asymmetric' carbon atoms – a carbon atom bonded to four different substituents – it will not be superimposable on its mirror image. Optical activity deriving from intrinsic molecular structure can occur in any state of matter: solid, liquid or vapour.

Even if the molecules themselves are not chiral, however, a substance may still manifest optical activity if – as in a crystal, for example – the molecules are arranged in a well-defined chiral structure such as that of a helix. The optical activity of crystalline quartz comes from the helical winding of achiral silicon dioxide molecules about the optic axis. If melted or dissolved in a solvent, such a substance would lose its optical activity[15]. Fused quartz, therefore, is not optically active.

It is worth noting that free atoms – i.e. atoms not bound in molecules nor subjected to static electromagnetic fields – are spherically symmetric and should have no preferential handedness. Correspondingly, the laws of electrodynamics (both classical and quantum) strictly forbid individual atoms from exhibiting optical activity. It turns out that they do, anyway – but about this I will comment in the next chapter.

In addition to optical rotation, another common manifestation of optical activity is that of circular dichroism, in which incident linearly polarised light is converted to elliptically polarised light by propagation through a light-absorbing chiral substance. Elliptical polarisation may be thought of as an unequal linear superposition of left and right circular polarisations. The phenomenon arises from the differential absorption – rather than phase shift – of the component left- and right-circularly polarised waves. An expression for the so-called ellipticity of the light – which, like optical rotation, is also expressible as an angle – would resemble equation (4.9) except that chirally asymmetric absorption coefficients ($\varkappa_L, \varkappa_R$) would replace the indices of refraction.

The two phenomena, optical rotation and circular dichroism, are, in fact, closely related. As illustrated in the previous section (relation (4.5a)), one can regard the index of refraction as a single complex number of which the real part characterises phase shifts, and the imaginary part characterises light absorption. Which phenomenon may occur for a given substance depends on the frequency (or wavelength) of the light. If the frequency corresponds to a Bohr transition frequency of the

---

[15] Liquid crystals, mentioned earlier, would actually constitute a separate category, for the molecules are neither perforce chiral, nor do they form a rigid chiral pattern. Rather, these optically anisotropic molecules are partially oriented within stacked horizontal layers with the direction of order turning progressively about the vertical axis from one layer to the next. The optical activity of liquid crystals falls outside the discussion of this chapter.

chiral system, the light is absorbed (presuming the transition is allowed) and circular dichroism results; if the frequency lies outside the regions of the spectrum where light is absorbed, then the substance is transparent and optical rotation occurs instead.

Besides modifications of the polarisation of *transmitted* light, there is yet another process by which a chiral material may manifest optical activity, namely light *reflection and refraction*. Although reflection and refraction may sound like two separate processes, they are basically dual manifestations of a particular example of light scattering at an interface. To derive the amplitudes of either the reflected or refracted waves, one must analyse both processes together. At this point it is appropriate to return to Fresnel and to enquire just how he tested his hypothesis of the existence of circularly polarised light.

There is a subtlety to the nature of circular polarisation not encountered with linear polarisation. Polaroid plastic, routinely found in sunglasses and commonly available from scientific supply houses, did not exist in Fresnel's day. Instead, birefringent materials such as Iceland spar were used to create and analyse linearly polarised light. An incident unpolarised light beam passing through a calcite crystal is decomposed into two transmitted beams with orthogonal linear polarisations. These two emerging beams would produce spots of equal brightness on some distant screen. If the incident light is linearly polarised, however, the resulting spots would be of unequal intensity depending upon the angle that the electric field of the incident light wave makes with the optic axis of the crystal. For an incident beam polarised either parallel to, or perpendicular to, the optic axis only one beam of corresponding polarisation would emerge from the crystal. That is, there would be only one spot on the screen.

Circularly polarised light passing through Iceland spar *also* gives rise to two spots of equal intensity, from which one might have erroneously inferred that the incident light was *un*polarised. Yet there is a world of difference between circularly polarised light, for which the electric field at any point in the wave rotates uniformly in one sense or the other about the direction of propagation, and unpolarised light in which case the orientation of the electric field fluctuates randomly and rapidly during the time the polarisation is observed. How is one to tell the difference?

Fresnel rightly understood that circularly polarised light may be thought of as a superposition of two linearly polarised light waves oscillating out of phase with one another by 90°. Indeed, he was able to create circularly polarised light by directing a linearly polarised light beam into a specially shaped glass prism – today called the Fresnel rhomb – and causing it to undergo total reflection at two opposing glass–air interfaces before emerging. With the electric field of the entering light oriented at 45° to the plane of incidence, the amplitudes of the reflected s- and p-polarised components incurred a relative phase shift of 45° for each total internal reflection. To show that the beam subsequently emerging into the air was circularly polarised, rather than unpolarised, Fresnel passed it again through a similar rhomb. After two additional total internal reflections, the resulting light (now with a phase shift of 180° between s- and p-polarised components) again became linearly polarised. By contrast, unpolarised light entering a Fresnel rhomb leaves as unpolarised light irrespective of the number of internal reflections.

Fresnel verified his interpretation of optical rotation – and correspondingly confirmed the existence of circularly polarised light – by means of adroit application of reflection and refraction within another of his ingenious prisms. If the refractive indices of right- and left-handed light are different in an optically active medium, then the two circular polarisations ought to refract at different angles. Thus, an initially linearly polarised light beam obliquely penetrating the surface of an optically active material should split into *two* beams of opposite circular polarisation. The expected effect, however, is extremely small. Yellow light (589 nm) from a sodium lamp, for example, propagating along the optic axis of a quartz crystal experiences a difference in refractive index, $|n_L - n_R|$, of about 7×10^{-5}.

With his customary inventiveness, Fresnel circumvented this difficulty by concatenating segments of left- and right-handed quartz prisms to make a composite prism. Linearly polarised light entered one end of the prism. At each interface between optically active quartz segments of opposite chirality, the deviation between refracting left and right circularly polarised light waves was enhanced. From the opposite end of the composite prism there emerged, as Fresnel predicted, two circularly polarised light beams sufficiently separated so as to leave no doubt about the existence of this 'new' form of light.

There is something both beautiful and ironic about Fresnel's researches on light reflection, polarisation and optical activity. I have often wondered whether Fresnel ever concerned himself with the problem of light reflection from an optically active medium. It would, after all, have been a natural thing for someone to do who derived the laws of reflection for ordinary (achiral) dielectrics and who employed the differential refraction of circularly polarised light to elucidate the nature of optical rotation. Nevertheless, I never found any mention of such an investigation in Fresnel's collected *Oeuvres*. In retrospect, this is perhaps not so surprising. Given the contemporary state of technology within which he had to work, any sought-for chiral effects could well have been impossibly weak to observe – even for a Fresnel. (Indeed, with the photometric methods available throughout the 1820s it was not possible to measure reflectance with sufficient precision to test the original Fresnel coefficients for achiral media.) What surprised me more, however, when I was just beginning to turn my attention to the study of optical activity, was to find that over a century and a half after Fresnel, this fundamental problem was apparently *still* not solved.

Although Fresnel's interpretation of optical rotation was fine as far as it went, it did not explain the origin of chiral refractive indices. How do left- and right-handed molecules or crystals specifically affect light differently? The answer to this question can be rigorously provided by a quantum mechanical description of the interaction of chiral systems with light. Nevertheless, the following more easily visualisable classical model provides a heuristic explanation that embodies the seminal features of the quantum treatment. Imagine a linearly polarised light wave incident upon an arbitrarily oriented helical molecule in a large sample of identical molecules (i.e. all the helices have the same handedness). Responding to the oscillating electric field, electrons in the molecule suffer periodic displacements about their equilibrium locations and give rise to an oscillating electric dipole moment. In addition, the alternating flow of electric charge along the helix engenders an oscillating magnetic dipole moment[16]. Similarly, the oscillating magnetic field of the light wave produces (by Faraday's law of induction) a time-varying electric field at the molecule that also gives rise to induced electric and magnetic dipole moments.

The key feature to note is that the relative orientation of the induced electric and magnetic moments depends only on the *sense* of the helical winding and not on the orientation of the helical axis. For example, the

[16] The separation of charges $+q$ and $-q$ by a distance d constitutes an electric dipole moment of magnitude qd directed from the negative charge to the positive charge. A current I around the periphery of a loop enclosing area A constitutes a magnetic dipole moment of magnitude IA/c (c is the speed of light) oriented perpendicular to the loop in a right-hand sense. (If the fingers of one's right hand encircle the loop in the direction of the current, then the extended thumb points in the direction of the magnetic moment.)

moments may be parallel for a right-handed helix and antiparallel for a left-handed helix. Oscillating electric and magnetic dipoles are themselves sources of electromagnetic radiation. However, the electric, and correspondingly the magnetic, fields of the waves emitted in a given direction by these two sources are perpendicular to one another. Thus, the waves superpose to yield a resultant scattered wave with electric and magnetic fields that are rotated with respect to the corresponding fields of the incident wave. The *extent* of rotation of the polarisation varies with the orientation of the helix, but the *direction* of the rotation depends only on the relative orientation of the induced dipole moments ... and this is the same for all the helices in the sample. Consequently, the net forward scattered wave, produced by the superposition of waves scattered from all the helices, is rotated clockwise or counterclockwise depending on the chirality of the helices.

Keeping in mind that the above description is only a model introduced for the purpose of helping make tangible what, in effect, is a lengthy mathematical analysis – and that it certainly does not account for all aspects of optical activity – one may still ask how the model accounts for 'circular birefringence', i.e. for different indices of refraction for the two forms of circularly polarised light. At no point in the discussion has a light wave been assumed to move at anything other than the vacuum speed of light. Why then are the refractive indices for right- and left-handed light different?

The *microscopic* description of light propagation through a transparent material refers mainly to relative phases: the relative phase of the induced dipoles, the relative phase of their radiated waves, the relative phase of the scattered and incident waves. It is only in the *macroscopic* or phenomenological description of optical activity, where light is presumed to interact with a continuum of matter, that the index of refraction is introduced as a means of accounting for the net phase retardation produced in a transmitted wave by all the molecules of the sample. Microscopically, there are only 'atoms (or molecules) and the void'; light moves at speed c in the interstices of matter only to be absorbed and re-radiated – and therefore apparently slowed – by molecular encounters. The net result is an effective lowering of the speed by an amount that depends, in the case of a chiral material, on the handedness of the molecules.

To recapitulate and generalise the salient points, the origin of optical activity derives from two distinct processes. The first, termed 'spatial dispersion', is the variation in the phase of an incident light wave over the extent of a chiral molecule or molecular aggregate. This is to be contrasted with the interaction of a light wave with an atom the 'size' of which (about 10^{-8} cm) is some three orders of magnitude smaller than the wavelength of visible light. To an incoming light wave, the atom is a mere point; the variation in phase of the wave over an atomic scatterer is usually neglible. The variation in phase of an incident light wave over a chiral molecule, however, reveals in the scattered wave the molecular handedness.

The second process is the interference that results from the superposition of scattered waves issuing from the electric and magnetic dipole moments induced in the medium by the incident wave. For elementary constituents that are not chiral, the incident light cannot induce electric and magnetic dipole moments simultaneously, and there would be no optical activity.

In the quantum mechanical treatment of optical activity, the role of symmetry is perhaps more direct and fundamental. The states of a quantum system with a centre of symmetry are characterised by sharp values of a quantum attribute termed 'parity'. If, upon inversion of the coordinates $(x,y,z \rightarrow -x, -y, -z)$ of all particles in the system, the wave function is unchanged, the state is said to have 'even' parity. If the wave function changes sign under coordinate inversion, the state is said to have 'odd' parity. An oscillating electric dipole can induce transitions between two states – and thereby produce light – only if the two states have *opposite* parities. By contrast, an oscillating magnetic dipole can induce transitions only between two states that have the *same* parity. Electric and magnetic dipole transitions, therefore, cannot occur simultaneously between states of sharp parity, and systems characterised by such states do not manifest optical activity. The quantum states of chiral systems, however, are superpositions of states of even and odd parity; an incident wave induces both electric and magnetic dipole moments as depicted in the classical heuristic model previously described.

From the phenomenological perspective of physical optics, the conceptual problem of describing optical activity can be considered resolved when the so-called constitutive, or material, relations are known. Although it suffices to speak simply of the 'electric' and 'magnetic' fields of a light wave in vacuum, the situation is more complicated for light in a medium. The electrons of the molecules, induced to oscillate by the electric and magnetic fields of the incident wave, serve as sources of additional internal electromagnetic fields. In all there are four types of electromagnetic fields generally designated E (the electric field), B (the magnetic induction), D (the electric displacement) and H (the magnetic field) whose properties one must know in order to predict the response of a material to light. The constitutive relations connect the secondary fields D and H (arising from induced charges and currents within the medium) to the fundamental fields B and E.

The constitutive relations for the simplest optically active medium – one that is intrinsically *non*magnetic, isotropic, homogeneous and transparent – were derived from quantum mechanics some sixty years ago[17]. I refer to it as the symmetric set of relations, because it has a form that remains invariant under a special type of symmetry transformation that effectively interchanges E and H, and B and D. However, an alternative and simpler set of relations – which I designate the asymmetric set – is the set one would most likely find in optics books that treat the subject of optical activity[18]. The chief feature of the asymmetric set is that the optical activity of the medium is presumed to derive exclusively from its dielectric properties; there is no magnetic effect of the light wave on the medium and, as a result, the magnetic fields B and H are effectively identical. Clearly, then, the above-mentioned symmetry transformation would not leave the theoretical expressions unchanged – hence the term 'asymmetric'.

Ironically, the two outwardly dissimilar sets of relations have both successfully accounted for optical rotation and circular dichroism in the transmission of light through an optically active medium. That is, when employed in Maxwell's equations, both sets predict the existence of circularly polarised waves with different refractive indices of the form

$$n_L = n_0(1+f), \tag{4.10a}$$

$$n_R = n_0(1-f), \tag{4.10b}$$

where n_0 is the mean refractive index, and the parameter f is a measure of the intrinsic strength of the chiral interaction between the medium and light. In the absence of evidence to the contrary, the symmetric and asymmetric sets of constitutive relations have long been considered physically equivalent. Indeed, this equivalence has been asserted as a consequence of another fundamental symmetry of the laws of classical electromagnetism. It has been argued that the fields D and H for a

---

[17] E. U. Condon, Theories of Optical Rotary Power, *Rev. Mod. Phys.* **9** (1937) 432.
[18] See, for example, G. R. Fowles, *Introduction to Modern Optics* (Holt, Rinehart and Winston, New York, 1975) 185–9.

particular medium (like the vector potential field *A* of the Aharonov–Bohm effect discussed in Chapter 1) are not unique – that it is always possible to transform a given pair to a new pair of fields *D'* and *H'* (that also satisfy Maxwell's equations) by redefining in a prescribed way the induced electric and magnetic dipoles. By means of one such family of transformations the magnetic dipoles can be made to vanish altogether in which case the optical properties of the medium derive only from the (redefined) electric dipoles. Such transformations effectively convert the symmetric set of constitutive relations into the asymmetric set apparently demonstrating that the different mathematical forms superficially mask a fundamental physical equivalence.

As I explained earlier, I first became interested in light reflection from optically active media as a possible means of selectively amplifying circularly polarised light. But other motivations rapidly developed as well. Ever since the discovery of natural optical activity, the principal experimental methods (optical rotation and circular dichroism) generally involve measurement of polarisation changes incurred by transmitted light. I was interested in exploring what new things might be learned by an alternative experimental technique. For example, the study of optical activity by light reflection could confer significant advantages in the investigation of chiral thin films which would be too thin to have much of an effect on a transmitted beam, or, conversely, in the investigation of opaque chiral samples through which a transmitted beam would be undetectably weak.

A more exotic potential application relates to the study of life itself. Biochemical processes carried out in the laboratory with substances of nonbiological origin ordinarily lead to equal mixtures of left- and right-handed molecular forms (called enantiomers) that display no residual optical activity. By contrast, the capacity to produce and consume optically active substances (sugars, amino acids, etc.) of a particular chirality is perhaps the most salient feature of life on Earth as we know it. Whether such chiral asymmetry in living things arose by chance or evolved deterministically from the laws of physics is not known. The question has far-reaching implications, however. A universal origin of biological homochirality would suggest that nonterrestrial life (if there is any) should display the same chiral preferences. It is conceivable that some day the manifestation of optical activity in the light reflected from planetary or asteroidal surfaces may signify the existence of living things.

Unaware of potential subtleties in the description of optical activity, I calculated the theoretical Fresnel coefficients and the resulting reflectances for incident light of linear and circular polarisations. To my surprise the analyses of light reflection based on the two 'equivalent' sets of material relations for an optically active medium gave entirely different results! Indeed, for light striking the surface of the material perpendicularly, the symmetric set of relations led to no differential reflection at all between left and right circularly polarised light – even though the reflecting medium has an intrinsic handedness. I viewed this result with considerable suspicion, and found more satisfying at first the prediction of the asymmetric set that at normal incidence occurred the largest differential reflection of circularly polarised light. This satisfaction was short lived. To my still greater surprise, the reflectance and transmittance predicted from the asymmetric set of relations did not sum to unity even for a transparent medium. These amplitudes violated the fundamental physical law of energy conservation!

Clearly, the two descriptions of optical activity were not equivalent (Figure 4.6). I wondered which set of Fresnel coefficients, if either, was correct. Was it possible that so basic a problem in the optics of chiral media could have gone unnoticed and untested since the development of classical electrodynamics a century and a half ago?

Untested it seemed to be, but not unnoticed. Several others had also been aware of theoretical inconsistencies in the amplitudes associated with the asymmetric set of material relations. But, as these relations were considered well established by previous studies of optical activity, the origin of the discrepancies was attributed to the structure of classical electrodynamics itself. Proposals were made to change the familiar Maxwellian expressions characterising the flow and conservation of energy or the boundary conditions of electromagnetic fields at interfaces.

It is one of the intriguing features of physics, perhaps of other sciences as well, that what is construed to be understood best turns out often enough not to be well understood at all. Then the importance of experimentation, too often forgotten or ignored as the elements of physics theory become increasingly abstract and remote from experience, must be reasserted. Maxwell, himself, expressed this sentiment eloquently in his 1871 Cambridge lecture celebrating the establishment of the course of Experimental Physics and the erection of the Devonshire Laboratory at Cambridge University:

This habit of recognising principles amid the endless variety of their action can

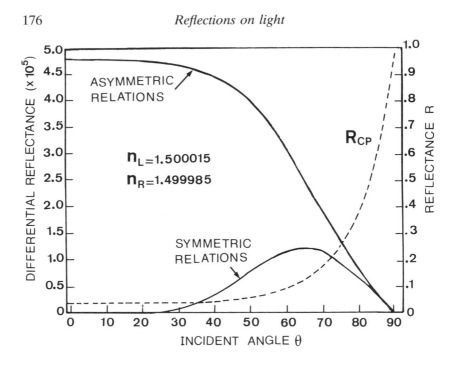

Figure 4.6. Theoretically predicted differential reflectance $D = (R_L - R_R)/(R_L + R_R)$ of left and right circularly polarised light from an isotropic optically active medium. The curve based on the symmetric constitutive relations shows a null D at normal incidence where the curve derived from the asymmetric constitutive relations is maximum. The individual reflection curves for left and right circular polarisations (dashed line designated R_{CP}) are not distinguishable on the scale shown at the right.

never degrade our sense of the sublimity of nature, or mar our enjoyment of her beauty. On the contrary, it tends to rescue our scientific ideas from that vague condition in which we too often leave them, buried among the other products of a lazy credulity . . .[19]

And so I was at the E.S.P.C.I. in Paris to test with my French colleague whether a widely accepted description of optical activity, or whether Maxwell's own electromagnetic theory (in its proper domain of classical optics), was one of the 'products of a lazy credulity'. The objective of our experiment was to measure the difference in the intensities of left-handed and right-handed light beams reflected by a naturally optically active sample. This difference is ordinarily very small, on the

[19] L. Campbell and W. Garnett, *The Life of James Clerk Maxwell* (MacMillan, London, 1984) 269.

order of the circular birefringence ($|n_L - n_R| \sim 10^{-5}$ or 10^{-6}) itself, and can be masked by a variety of instrumental artefacts. This experiment was to be a difficult one.

One might wonder why a reflection experiment should give rise to so weak an effect, when a transmission experiment can lead to a readily measurable optical rotation. The answer lies in part in equation (4.9). Besides the circular birefringence, the expression also contains the ratio of the sample thickness to the wavelength of light. By employing an optical path length through the sample many times larger than the wavelength of light, the intrinsically weak circular birefringence can be effectively amplified. The differential reflection of left and right circularly polarised light from the surface of a bulk (ideally infinitely thick) optically active medium is insensitive to the thickness of the medium and does not depend on the wavelength explicitly (although there is an implicit dependence through the index of refraction). Thus, it cannot be amplified in this way.

Nevertheless, there are ways of enhancing the difference in reflected circularly polarised light. One way is by 'index matching' or adjusting the mean refractive index ($n_2 = \frac{1}{2}(n_L + n_R)$) of the optically active medium to be close to that of the achiral medium (n_1) within which the light originates. As a rough approximation, the difference in reflectances is

$$R_L - R_R \propto \frac{|n_L - n_R|}{[(n_2/n_1)^2 - 1]} \qquad (4.11)$$

for a transparent medium with $n_2 > n_1$. There comes a point of diminishing returns, however, for the closer the indices are matched, the greater is the transmitted light, and the weaker is the reflected light – unless the material is absorbing. In the latter case, the refractive index is complex (equation (4.5a)), and matching the real part to the index of the achiral medium does not lead to a vanishingly small reflectance. A second way employs multiple reflection of light at the optically active surface. Under appropriate conditions – again involving an absorbing medium – the difference in reflectances for circularly polarised light can be made to increase linearly with the number of reflections. Both these methods were eventually to play a significant role in our experiments.

The instrumental heart of the reflection experiment is the photoelastic modulator (PEM), which makes it possible to determine the difference in reflectance of circularly polarised light nearly instantaneously in a single-step measurement. Since this difference is small compared with intensity fluctuations of the light source, it would be a hopeless endeavour to attempt to measure the reflectance of left and right circularly polarised light separately and then to substract them. Stripped to its bare essentials, the PEM is a bar of fused (and therefore optically inactive) quartz made to oscillate at a frequency of about 50 kilohertz (kHz) along its long axis (Figure 4.7). A light beam, before or after reflection from the surface of an optically active sample, traverses the quartz bar in a direction perpendicular to the axis of oscillation. As a result of the mechanical vibration, the refractive index for light polarised along the axis of the bar also oscillates at 50 kHz. Thus, upon passing through a PEM vibrating at frequency f, the light itself incurs an oscillatory phase shift of the form

$$\phi = \phi_0 \sin(2\pi f t) \tag{4.12a}$$

between the components of the beam polarised parallel to, and perpendicular to, the axis of the quartz element.

The effect that the modulator has on the light depends on the modulation amplitude ϕ_0 and on the polarisation of the incident light. Suppose the incident light is linearly polarised at 45° to the vibration axis of the quartz. The parallel and perpendicular components of the wave (of frequency v) can then be expressed as

$$E_{\parallel} = \cos(2\pi v t - \phi), \tag{4.12b}$$

$$E_{\perp} = \cos(2\pi v t). \tag{4.12c}$$

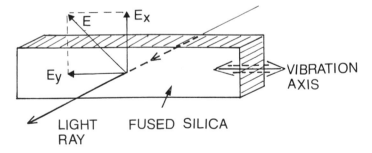

Figure 4.7. Schematic diagram of the photoelastic modulator (PEM), a bar of fused silica made to vibrate at an ultrasonic frequency along its long dimension. Periodic compression and extension of the intrinsically isotropic bar makes the bar birefringent with a refractive index oscillating at the same frequency along the axis of mechanical vibration. The electric field E of a light wave passing transversely through the PEM is a superposition of components (E_x, E_y) linearly polarised perpendicular and parallel to the vibration axis. The PEM produces an osillating relative phase between these two components.

With ϕ_0 set equal to $\pi/2$ radians, the emergent wave oscillates harmonically between left and right circular polarisations as the quartz bar is maximally extended and compressed at the mechanical frequency of 50 kHz. At intermediate positions of the end of the bar, the wave polarisation is elliptical, i.e. a linear superposition of left and right circularly polarised components. Other settings of the modulation amplitude produce emergent waves that evolve in time between other states of polarisation. For example, if ϕ_0 is set to π radians, then the transmitted wave oscillates harmonically between two orthogonal states of linear polarisation. Linearly polarised light, as Fresnel first showed, is also resolvable into a superposition of circularly polarised components. In effect, the PEM makes it possible for the chiral reflector to sample incident left and right circularly polarised light of equal intensity at least once every period of mechanical oscillation ($T = 1/(50\,\text{kHz}) = 0.02$ millisecond), a time scale short compared with that of light source fluctuations.

Received by a photodetector, the modulated light gives rise to an electric current in the output containing a d.c. component (0 Hz) as well as components oscillating at the fundamental frequency (50 kHz) and higher harmonics (100 kHz, 150 kHz, etc.). The various components can be measured individually by means of an instrument known as a lock-in amplifier or synchronous detector. A quantitative measure of the differential reflection of circularly polarised light can then be determined directly from appropriate ratios of these current components.

In principle, at least, that is how the experiment was supposed to work. Were matters actually so simple, the desired data could have been collected within a week. In reality, however, the experiment, first begun in the mid-1980s with my students, was pursued over a period of years as it became necessary to find and eliminate spurious signals that mimicked the effects of optical activity. One of the last and trickiest problems stemmed from the PEM itself. Although reliably employed since the mid-1960s to measure the circular dichroism of optically active materials with a sensitivity of one part in a hundred thousand, the PEM now appeared to give rise to a curious false signal two orders of magnitude larger than ever expected when used to test the chiral Fresnel coefficients. Nor was this a 'local' phenomenon; the signal persisted for all PEMs tried, whether of commerical origin or home-made. How was it possible for scores of optical physicists using similar devices over a period of more than two decades to have missed so large a systematic error?

It was not possible; this artefact, as Jacques and I were to understand better later, effectively did not show up in experiments employing only one polariser. In our experiments, however, there were always at least two polarisers: the one that prepared the light beam for passage through the modulator, and the reflecting surface itself (the 'hidden' polariser of my kitchen experiment years earlier).

The solution to the problem, which was to have an intrinsic utility of its own, was traced to the existence of a weak static *linear* birefringence in the quartz rod. This is the type of optical anisotropy found, for example, in calcite where linearly polarised waves pass through the crystal at different speeds depending on their direction of propagation and orientation. Ironically, it had long been known that stresses induced by the manufacturing process or by pressure from the edges upon which the bar rested generated a weak linear birefringence in the quartz. It was always assumed, however, that the axis of static birefringence lay parallel to the axis along which the quartz oscillated. In fact, we had expressly designed at the outset an experimental configuration for which this type of birefringence should not affect the desired signal.

The chiral reflection experiments showed that the axes of static and oscillating birefringence in the fused quartz were not parallel, and that even minuscule angular deviations – which ordinarily would have been inconsequential in measurements of circular dichroism – yielded disturbingly large signals in other experimental configurations. How was one to circumvent a problem that seemed to be an unavoidable consequence of fabricating the most essential part of the experiment! After all, the PEM still remained the most suitable way (short of counting individual photons – a procedure that was not without its own difficulties) of observing weak chiral asymmetries in scattered light.

The reflection experiment was temporarily put aside in order to examine in minute detail, both theoretically and experimentally, the passage of light through an elastic modulator. The comprehensive theory that finally emerged from this detour happily suggested a number of ways to circumvent the stress-induced birefringence of the quartz, if not also to reduce it, at least in the small region through which the light passes.

The experiment, back on track, is at the time of this writing still in progress. By taking advantage of index matching and multiple reflection, we have been able to observe for the first time the difference with which a naturally optically active material reflects left- and right-handed light (Figure 4.8). It is a testament to the extraordinary experimental

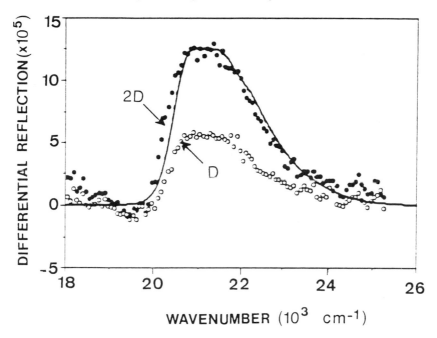

Figure 4.8. Experimentally observed differential reflection of circularly polarised light from an optically active liquid (camphorquinone in methanol) as a function of wave number (reciprocal wavelength which is proportional to photon energy) at an incident angle of 67°. D denotes single reflection and 2D denotes double reflection from the chiral medium. The solid line shows the theoretically calculated curve.

ability of Augustin Fresnel that the complementary phenomenon to the differential circular refraction of light has only now, after more than one hundred and seventy years, been achievable. As to the theoretical description of optical activity, although final results are not in, preliminary indications are that neither Maxwell's electrodynamics nor quantum theory is likely to be embarrassed. Rather, a number of optics books could well stand some revising. In my own mind I have no doubt that the symmetric description of optical activity is the correct one.

But what about the argument purportedly demonstrating the equivalence of the two sets of chiral material relations? The argument fails. Although it is indeed true that the prescribed transformation connects the two phenomenological descriptions of optical activity while leaving Maxwell's field equations unchanged, the transformation does *not* leave the Maxwellian boundary conditions unchanged. If the symmetric set of relations is correct, then use of the asymmetric set together with standard electrodynamics constitutes a theoretically inconsistent calculation. No wonder that the resulting amplitudes violate physical laws.

The episode reminds me somewhat of the theoretical arguments against the Aharonov–Bohm effect (Chapter 1) in which case the vector potential of the confined magnetic field could allegedly, but not actually, be transformed away. In that case, the particular gauge transformation was not legitimate. In the present case, the symmetry transformation, while not disallowed, had been applied to only part of the mathematical framework needed to solve the problem of reflection.

If the amplitudes deriving from the symmetrical material relations are correct, how is one to understand why an intrinsically chiral material reflects left- and right-handed light *equally* at normal incidence? Why is the intrinsic chirality of the medium not manifest? A heuristic explanation of this puzzling behaviour may be sought again in the microscopic model of reflection justified by the Ewald–Oseen extinction theorem discussed in the previous section.

Consider first the passage of light through a transparent optically active material, a medium with no distinguishing optical axes. The sense of optical rotation is determined exclusively with respect to the direction of propagation. Suppose a linearly polarised wave propagates ten centimetres from right to left during which passage the plane of polarisation is rotated 45° towards the left-hand side of someone looking at the light source. Let the wave then be reflected and made to propagate ten centimetres back again from left to right. The plane of polarisation is rotated an additional 45° towards the left-hand side of an observer looking towards the light source – but in this case the light source is the reflecting mirror, and the second observer is facing the direction *opposite* that of the first observer. Thus, the plane of polarisation is actually brought back exactly to its original orientation. In other words, the net optical rotation of a light beam that has made an even number of passes back and forth through a naturally optically active medium is zero.

In regard to the problem of chiral reflection, an incident light beam does not interact with the reflecting medium at the surface only, but may be thought of as propagating into the medium, being absorbed and thereby inducing molecular dipoles to radiate secondary waves that superpose coherently to form the reflected wave. At normal incidence this interaction is equivalent to a penetration of the wave into the medium followed by reflection as if from a perfect mirror. Upon reflection and propagation back to the surface the net chiral effect vanishes. At all other angles of incidence, except for grazing incidence (where the differential reflection of circularly polarised light is again zero), the planes of optical rotation of the incident and reflected waves are no longer parallel and exact cancellation does not occur.

It is worth noting in conclusion that, with a self-consistent theory of chiral reflection at hand, it was possible to return to the question that sparked my study of optical activity in the first place: can circularly polarised light be selectively amplified by reflection from an optically active medium with a population inversion? The answer is yes, and possibly one day this process may provide a new way to probe the chemical structure and physical interactions of excited molecules or prove useful in devices to amplify light. Hopefully, the first experiment will not turn out to be another Maxwellian demon.

* * * * * * * * * * * *

Though the optically active demon has yet to be exorcised definitively, the curious controversy over left- and right-handed light reflection has already produced results of conceptual and practical interest to scientists and engineers concerned with the origin and measurement of small chiral asymmetries in matter. But one theoretical study in particular has emerged in the course of this research that addresses a most unusual interaction.

Years ago, when I was first introduced to the quantum mechanics of atoms and molecules, I often wondered whether there may be an unsuspected structure to atomic energy levels finer even than the finest structure described in the textbooks – finer than the structure due to the magnetic coupling of electrons and protons or to the interactions of electrons with the 'vacuum'. My interests have ranged widely since that time, but old questions often have a way of returning until they are answered. One day I found an answer, and it quite literally lay right under my feet.

5

Two worlds, large and small: Earth and atom

As I stood in an unlighted horizontal passageway of the Cashmere Cavern and looked up at the narrow ventilation shaft receding to a small circular opening some twenty metres above me, I felt a shiver of fascination and amusement as I thought of my New Zealand colleagues being raised and lowered by a rope harness. Located under the Cashmere Hills just outside Christchurch on the South Island, the cavern was originally excavated about a half century ago by the military to serve as a World War II command post in the event of a Japanese invasion. With the passage of time it had long since faded from public memory until rediscovered recently by accident. A timely rediscovery, too. With its solid bedrock floor and sheltered environment, the cavern is expected to provide an ideal workplace thirty metres below ground for the newly created University of Canterbury Ring Laser Laboratory. A wide horizontal adit seventy metres in length now gives easy, if less dramatic, access for the construction crew and physicists.

During the summer – i.e. Southern Hemisphere winter – of 1990, while at the University of Canterbury to deliver a series of lectures, I observed the ongoing construction of the new laboratory and testing of the laser system with more interest than merely that of a curious visitor. When completed, this facility may quite possibly be able to confirm a remarkable optical effect that has long interested me: that unbound atoms unperturbed by static electric or magnetic fields on the rotating Earth interact inequivalently with left and right circularly polarised light.

Actually, it has been known for over thirty years that atoms can be optically active as a result of the weak nuclear interactions[1]. These

[1] F. Curtis Michel, Neutral Weak Interaction Currents, *Phys. Rev.* **B138** (1963) 408.

interactions are not invariant to reflection in a mirror and therefore can be expected to engender a left–right asymmetry in the quantum states of atomic electrons. That is, if they did not destroy the integrity of the atom, for the weak interactions are usually associated with particle disintegration processes as in the familiar example of beta decay, the natural transformation of a neutron into a proton, electron and antineutrino.

In order for the weak interaction to break the chiral symmetry of bound electron states, without at the same time altering the identity of the atom through some charge-changing process, there must be a way for electrons and nucleons (the constituents of the atomic nucleus) to interact by exchange of a massive neutral particle. Just such an interaction is provided by the so-called 'electroweak' theory, a sweeping theoretical synthesis of electrodynamics and the weak interactions unmatched in scope since Maxwell unified all of electricity, magnetism and optics. Within the framework of this theory electrons and nucleons can exchange a Z^0 vector boson, a neutral particle with a mass approximately one hundred times the mass of a proton. The existence of such an exchange or 'weak neutral current' was demonstrated in 1973 by high-energy experiments involving the scattering of neutrinos by nucleons.

The corresponding existence of atomic optical activity was confirmed in the early 1980s by low-energy experiments on the vapours of a variety of heavy atoms such as bismuth, lead and thallium[2]. The effect is small; one metre of dense vapour can rotate the plane of linear polarisation of a transmitted light beam by about 10^{-7} radians.

The atomic optical activity that I predicted, however, has nothing whatever to do with the weak interactions. It arises, instead, from the rotation of the Earth and is many times weaker than any which has heretofore been measured. The very existence of this phenomenon, however, captures the imagination. For one thing, the weak interactions aside, the laws of electrodynamics exhibit perfect mirror symmetry from which it follows that optical activity in free atoms – spherically symmetric systems held together by an electrostatic force – is ordinarily strictly forbidden. Secondly, apart from the issue of atomic handedness, the predicted effect represents a potentially observable influence of planetary spin on the internal workings of an atom.

* * * * * * * * * * * *

[2] M. A. Bouchiat and L. Pottier, Optical Experiments and Weak Interactions, *Science* **234** (1986) 1203.

It is the essential dichotomy in the application of the laws of physics to systems large and small that makes the thought of such an atom–planet interaction so unusual. To predict the orbit of a comet about the Sun or of the Moon about the Earth, one relies on the laws of classical mechanics as embodied in Newton's equations of motion. Correspondingly, to determine the motion of an electron about a nucleus – or in other words to understand the structure of the atoms and molecules out of which the objects of the macroscopic world are built – one turns instead to the laws of quantum mechanics as embodied in entirely different equations, for example those of Schrödinger, Heisenberg and Dirac. This recourse to separate and incommensurate theoretical frameworks for deducing the behaviour of large-scale and ultra small-scale objects reflects in a profound way the decoupling of the objects themselves. The motion of a single atom hardly influences the daily affairs of a planet; likewise the motion of a planet ordinarily has no perceptible effect on the internal dynamics of an atom. It is to the extreme smallness of Planck's constant ($h = 6.67 \times 10^{-27}$ erg-second) that one may effectively attribute this decoupling of the large and small. A fortunate circumstance, too. Were Planck's constant much larger so that the fall of an apple could be accounted for only in the counterintuitive terms of quantum physics, one might well wonder whether even an Isaac Newton could ever have made sense of the world[3].

Nevertheless, the decoupling of the very large and the very small is not total. For one thing, as may be expected, even the elementary particles are influenced by the gravitational force of the Earth for, after all, bulk matter is made up of protons, neutrons and electrons. The manner in which they are affected, however, has given rise to some surprises. It has been known for centuries that a bulk object of mass m near the Earth's surface is attracted towards the centre of the Earth by a force of magnitude mg, where g is the local gravitational acceleration ($g \sim 9.8$ metres/s^2). Although individual neutrons fall freely in this expected way, experiments to probe the free-fall of electrons[4] through an evacuated vertical cylindrical tube of copper showed that the force of attraction was less than 10% of mg. Are electrons exempt from the law

---

[3] An exposition of the exaggerated relativistic and quantum phenomena to be expected if the speed of light were much smaller, and Planck's constant much larger, than their present values may be found in George Gamow's delightful Mr Tompkins books combined in *Mr. Tompkins in Paperback* (Cambridge University Press, London, 1971).

[4] F. C. Witteborn and W. M. Fairbank, Experimental Comparison of the Gravitational Force on Freely Falling Electrons and Metallic Electrons, *Phys. Rev. Lett.* **19** (1967) 1049.

of gravity? Actually, the result was no violation of Newton's law of gravity; on the contrary, it confirmed it in an unusual context.

A metal can be regarded in some ways as a rigid ionic lattice permeated by a mobile electron gas. At moderate temperatures and in the absence of external perturbations attractive electrostatic forces prevent the electrons from escaping from the metal surface, but within the metal interior loosely bound (valence) electrons are free to circulate. Before the electron free-fall experiment was performed, two theorists, Schiff and Barnhill, realised that the mobile electrons within a metal should also fall vertically in response to the pull of gravity[5]. The descent terminates, however, when the downward pull of gravity is balanced by the upward electrostatic attraction of the positive ions. The net downward displacement of the negatively charged electrons relative to the positively charged metal lattice creates an electric field directed downward *outside* the metal surface. This electric field exerts an upward force on electrons falling freely through the copper tube and thereby retards their acceleration. Indeed, if the gravitational acceleration of an electron within the metal is the same as that of a free-falling electron in empty space, then the magnitude of the electrostatic field outside the metal surface should be about mg/e (where e is the electron charge). This field would in principle counterbalance the pull of gravity on a free-falling electron, which, if dropped down a vertical metal tube, should then not fall at all! With account taken of nonidealities in the metal surface, this is effectively what was observed.

It is of interest to note that if a positron, the positively charged antiparticle of an electron, were introduced into the tube, the force of gravity and the Schiff–Barnhill force would now both be pulling downward. Thus, a positron would be expected to fall through the tube at *twice* the acceleration of gravity. I am not aware that this experiment has ever been done. However, had such an effect been observed in ignorance of the Schiff–Barnhill effect, the apparently nonsymmetrical action of gravity on particles and antiparticles would doubtless have created much excitement within the physics community, for it would seem to violate Einstein's equivalence principle, one of the seminal principles upon which the present understanding of gravity is based. In fact, recently reported anomalies in the interaction of the Earth with normal matter (rather than with antimatter) appear to manifest just such a violation.

[5] L. I. Schiff and M. V. Barnhill, Gravitational-Induced Electric Field Near a Metal, *Phys. Rev.* **151** (1966) 1067.

In effect, one version of the equivalence principle (of which there are various inequivalent versions) maintains that mass, alone of all the conceivable attributes of matter, determines the force of gravity that an object experiences. Actually, there are two conceptually different types of mass. One is inertial mass appearing in the definition of linear momentum (mass × velocity) and consequently in Newton's second law of motion

$$\text{Force} = \text{Inertial Mass} \times \text{Acceleration} = m_I a. \tag{5.1a}$$

The other is the gravitational mass introduced in Newton's law of gravity, a special case of which is the familiar expression

$$\text{Force} = \text{Gravitational Mass} \times \text{Free-fall Acceleration} = m_G g \tag{5.1b}$$

for the force of gravity near the Earth's surface. The so-called weak principle of equivalence affirms as an exact identity the experimentally observed numerical coincidence of the inertial and gravitational masses. It would then follow from relations (5.1a) and (5.1b) that, if m_I and m_G were always equal, two lumps of matter should fall freely at the same acceleration in response to the pull of gravity irrespective of differences in mass, isotopic composition, chemical structure or physical state (e.g. solid or liquid). Although a particle and its antiparticle may differ with respect to such properties as the sign (but not magnitude) of electrical charge or the relative orientation of spin angular momentum and magnetic moment, they are believed to have exactly the same mass and therefore, if the equivalence principle is valid, to behave identically in a gravitational field.

Galileo is alleged to have been the first to test the equivalence principle by dropping different objects from the Leaning Tower of Pisa, although it is questionable whether he really performed such an experiment. Credit for the first precision tests is usually accorded instead to Baron Roland von Eötvös of Hungary, whose series of experiments begun in the late 1880s and completed by 1922 remained the state of the art until the early 1960s[6]. Eötvös constructed a torsion balance – a device in which two masses of different composition were suspended at opposite ends of a horizontal bar supported at the centre by a thin fibre about which the bar could turn. Because of the Earth's rotation, each mass is subjected not only to the vertical pull of gravity, but also to the

[6] R. von Eötvös, D. Pekar and E. Fekete, Beiträge zum Gesetze der Proportionalität von Trägheit und Gravität, *Ann. Phys. (Leipzig)* **68** (1922) 11.

centrifugal force which acclerates the mass outwardly in a direction perpendicular to the axis of the rotation. Although sometimes designated a fictitious or pseudo force, since it originates in the acceleration of the reference frame, the centrifugal force gives rise to real enough physical consequences as judged by a co-rotating observer. Anyone who has felt himself thrown outwards as he drove a motor car round a turn in the road has experienced the centrifugal force. The centrifugal force on an object constrained to rotate with the Earth is proportional to the inertial mass of the object, the perpendicular distance of the object from the rotation axis, and the square of the angular velocity of rotation.

Concerning the Eötvös balance, if the ratio of the inertial to gravitational mass were different for each of the two suspended masses – that is, if the centrifugal force acted on one object proportionately greater than did the force of gravity – there would be a torque (i.e. a twisting effect) on the rod causing it to rotate about, and therefore to twist, the fibre. The small angle through which a torsion fibre is twisted can be measured to high precision, for example by an optical technique whereby an incident light beam is reflected from a small mirror affixed to the fibre. The angle of reflection, which is twice the angle through which the fibre is twisted, may be small, but the linear deviation of the reflected light beam from the incident direction increases in proportion to the distance between the detector and the mirror, and can be made measurably large. However, since one cannot turn off the rotation of the Earth, the equilibrium position of the balance arm provides no information about the relative influence of gravity and inertia; one cannot tell what the orientation would have been if gravity, alone, acted on the masses.

The key point to recognise is that, in the event that the equivalence principle is violated, the equilibrium orientation would depend on which mass is located on which side of the balance. Suppose the torque on the fibre for a given configuration of the masses orients the balance at equilibrium along an east–west directed line. Exchanging the two masses or, equivalently, rotating the entire apparatus (including the frame to which the fibre is mounted) by 180° would then cause the fibre to twist in the opposite sense, and the balance would no longer lie along the east–west line. If gravitational and inertial mass were truly identical, there would be *no* differential torque, and the balance would maintain the same orientation irrespective of the side on which each mass was located. By searching for such a change in equilibrium orientation,

Eötvös was able to establish the equivalence of gravitational and inertial mass for a number of dissimilar materials – readily recognisable ones like copper, water and platinum, as well as puzzling oddities like 'snake-wood' – to a few parts in a billion. Subsequent tests by other researchers in which Eötvös's basic procedure was implemented in a torsion balance that responded to the gravitational force of the Sun and the centrifugal force of the Earth's motion around the Sun established the equivalence of inertial and gravitational mass to precisions some three orders of magnitude beyond what Eötvös obtained.

Ironically, a re-examination in 1986 of Eötvös's definitive paper of 1922, sparked a lively controversy when the examiners concluded that, contrary to the long-held interpretation, the data in the paper actually provided evidence for a composition dependence of the gravitational acceleration[7]. The origin of this effect (the establishment of which is far from certain) has been attributed to an attractive 'fifth force', a new fundamental interaction (complementing gravity, electromagnetism and the strong and weak nuclear interactions) that depends not just on total mass, but on certain properties of the 'heavy' elementary particles (the baryons) of which a mass is composed. Protons and neutrons are the principal baryons composing ordinary matter. In the contemporary theory of elementary particles, each baryon is ascribed quantum numbers with whimsical names like baryon number, isospin, hypercharge or strangeness that play an important role in the various interactions resulting in particle transformations. The origin and significance of these numbers are of no concern here, but it is relevant to note that the baryon number for protons and neutrons is +1, the number for the corresponding antiparticles (antiproton and antineutron) is − 1, and that the net baryon number of a sample of ordinary matter is just the sum of the protons and neutrons.

The examiners of Eötvös's paper claim to have found that recorded differences in the accelerations of two masses – which ideally ought to be zero, but which, of course, like all experimental data, exhibit uncertainties due to the limitations of measurement – are not statistically random, but correlate with differences in the baryon number per unit mass of the sample. Although the baryon number is the same for all the baryons that make up ordinary matter, the baryon number per unit of mass is not necessarily the same for dissimilar materials, since the packing of the baryons can be different. One finds that the baryon

---

[7] E. Fischbach *et al.*, Reanalysis of the Eötvös Experiment, *Phys. Rev. Lett.* **56** (1986) 3.

density is greatest for elements around iron in the middle of the periodic table than for elements at either end. Thus, if the fifth force exists, the net interaction between the Earth and a certain mass may depend on whether that mass comprises more protons than neutrons or more neutrons than protons, or even whether it is made of antiprotons and antineutrons rather than of protons and neutrons.

That the postulated fifth force may have eluded physicists for so long can be explained in part by its intermediate range of action estimated to be a few tens to hundreds of metres – a distance scale enormous in comparison with that over which nuclear forces prevail (on the order of 10^{-13} cm) and negligible in comparison with the supposedly infinite range of gravity. Numerous studies of particle collisions in high-energy accelerators have yielded information about the interactions within and between nuclei. Correspondingly, astronomical observations of objects within and beyond the Solar System have long probed the effects of gravity. But virtually no experiments were designed in the past to test specifically for the existence of an intermediate-range interaction between matter. Actually, subtle manifestations of the fifth force may have already appeared in high-energy experiments with a peculiar family of particles (the K mesons), as well as in discrepancies between satellite and terrestrial measurements of the local gravitational acceleration g.

As one can well imagine, the prospect of finding a new force in nature was bound to stimulate a flurry of new experiments. Unfortunately, in the aggregate the results of recent efforts to detect the fifth force are contradictory and inconclusive with some experiments leading to positive results and others to null results. Although it is difficult to know for sure what lies at the root of the discrepancies, one obvious possibility, given that all experiments were performed terrestrially, is the unaccountable influence of near-by masses. Indeed, a number of the experiments depended on the presence of naturally occurring large concentrations of mass (like cliffs or mountains) to produce a differential effect on suspended test masses.

One approach, different from any that has yet been tried, occurred to me shortly after the controversy first began; it was to search for a deviation from Newton's law of gravity by means of a satellite experiment. The basic principle exploits a well-known, but nonetheless extraordinary, property of any force whose magnitude diminishes as the inverse square of the distance from its source (in this case a point mass). This attribute is shared by both the Coulomb force and (to the extent that general relativistic effects can be neglected) the force of gravity. In his *Principia*, Newton demonstrated mathematically that a test object outside a spherical distribution of mass is gravitationally attracted as if all the matter of the sphere were concentrated at the centre. Suppose, however, the sphere were hollow – a shell rather than a solid sphere – and the test mass lay *inside*. With what gravitational force would it be attracted to the walls of the shell?

Clearly, on the basis of symmetry alone, one could see that no force at all acts on a test mass at the centre of the shell. All directions leading away from the centre are equivalent; there can be no preferred direction of acceleration. What is perhaps less obvious is that the net gravitational attraction of the test mass by the shell is null *everywhere* in the shell interior! Newton understood this, too. The result holds for an arbitrarily thick shell and for a test mass of arbitrary shape and size, as long as it is entirely contained within the cavity of the shell.

The vanishing of the gravitational force within a spherical shell may be understood heuristically in the following way. Suppose the test mass is an ideal mass point (one of the most frequently found items in the physicist's stockroom of imaginary objects), and the shell is very thin. Extend a straight line drawn through the test mass in both directions until it intersects the shell at two locations. Unless the mass is at the centre of the shell, in which case we already know that it experiences no net gravitational force, one point of intersection is closer to the test mass than the other. Move the line (keeping the test mass fixed) so that each of the two segments generates the shape of a narrow cone with a circular base traced out on the inside surface of the shell. Consider the gravitational force exerted on the test mass by just those two portions of the surrounding shell contained within the circular bases.

The gravitational force exerted by a minute chunk of mass (let us call it an atom although the argument does not depend in any way on the discreteness of matter) in the closer section is greater than the corresponding force exerted by an atom in the farther section. On the other hand, the surface area of the farther section is larger in proportion to the square of the distance from the test mass and therefore contains more atoms. For the case of an inverse-square force law *only*, the stronger attraction by the atoms of the nearer region is exactly counterbalanced by the greater number of attractors of the more distant region with the result that there is no net gravitational force on the test mass from those two sections of shell.

Since the orientation of the line originally drawn through the test mass (i.e. the generator of the two cones) is entirely arbitrary, the net force on the test mass from *any* two sections of shell so delineated will cancel. In their entirety, all such mass sections constitute the whole of the thin spherical shell which therefore exerts no net force on the point mass inside regardless of its location. And since this conclusion holds for a thin spherical shell of any radius, it must be valid as well for any number of *concentric* thin shells, or equivalently for a single shell of arbitrary thickness. Furthermore, the net force on a test mass of *finite* size contained within the shell must also vanish if each point mass of which it is composed experiences no force.

The spatial dependence of the suspected fifth force is not purely inverse square, but is thought to diminish exponentially with distance. To express the mathematical form of a fundamental interaction, it is often more convenient (and sometimes absolutely necessary) to consider energy rather than force. The potential energy of two point masses, m_1 and m_2, separated by a distance r and interacting through both gravity and the fifth force, may be represented as follows

$$U(r) = -(Gm_1m_2/r)[1+\beta\exp(-r/b)]. \qquad (5.2a)$$

Newton's universal constant of gravity $G \sim 6.7\times10^{-11}\,\text{N}/(\text{m}^2\,\text{kg}^2)$ sets the scale of intrinsic strength of the gravitational interaction, whereas β is a dimensionless coupling constant that sets the corresponding scale of strength (relative to gravity) of the fifth force of which the characteristic range is b. On the basis of both geophysical data and reanalysis of Eötvös's paper, the coupling and range parameters have been estimated to be $\beta \sim -(7.2\pm3.6)\times10^{-3}$ and $b \sim 200\pm50$ metres. A negative β implies that the fifth force is attractive. To determine the actual force one mass exerts on the other, one must calculate the negative derivative of $U(r)$ with respect to r. The resulting expression, which need not be reproduced here, clearly gives $1/r^2$ dependence in the special case when β is zero (no fifth force). One can employ relation (5.2a) to determine the total potential energy of a test mass m_1 inside a spherical shell constructed from an infinite number of infinitesimal units of mass m_2. As before, the force is then calculable from $-dU(r)/dr$. Again, the resulting expression is somewhat cumbersome, but the principal result is easy to state: when β is *not* zero, the net force on a test mass inside the shell does *not* vanish.

To understand why, consider again the special case of a point mass in a spherical shell. As a result of the exponential factor $\exp(-r/b)$, the forces exerted by the two patches of shell formed by conical projections from the test mass no longer cancel. The force of each patch diminishes with distance to a greater extent than the patch area increases; the mass patch closest to the test mass therefore exerts a greater force than does the more distant patch.

An interesting consequence of this, which follows from relation (5.2a) if the fifth force is attractive, is that a test mass located anywhere within the shell (for a shell size small compared with the range b) will be pulled towards the centre with a strength of attraction linearly proportional to its displacement from the centre. This is the type of restorative force, referred to as Hooke's law, which gives rise to periodic motion about a point of equilibrium. The frequency f with which a test object of inertial mass m_I and gravitational mass m_G would oscillate within a spherical shell of inner and outer radii R_1 and R_2, respectively, and mass density μ can be shown to be

$$f = (2\pi b)^{-1}[(4\pi/3)G(m_G/m_I)|\beta|\mu(R_2^2 - R_1^2)]^{\frac{1}{2}}. \qquad (5.2b)$$

Thus, the occurrence of a harmonic oscillation at a frequency proportional to the square root of $|\beta|$ and inversely proportional to b, where no Newtonian gravitational force would be expected at all, would be an experimental signature of the putative new interaction.

The above considerations apply, of course, in the absence of forces from outside the shell. Here is where a satellite could prove useful. Orbiting around the Earth – or some other parent body – the spherical shell and all its contents are in a permanent state of free-fall. To a first approximation, therefore, the gravitational influence of the entire Earth has been eliminated. Like the astronauts in an orbiting space station, a test mass within such a shell would be weightless, its motion relative to the shell deriving ideally from those interactions that deviate from an inverse square spatial dependence. Moreover, the contribution of the fifth force of the planet should be negligibly small for a satellite located at an orbital radius many times larger than the range of the force.

If, as in the Eötvös experiment, one employs *two* test objects of different composition, i.e. differing in the proportion of baryons to inertial mass, the objects will oscillate about the centre at different frequencies. The possible advantage of a satellite experiment may then be seen in the following. Compared with the fractional difference in acceleration $\Delta a/g$ that these two masses would undergo in a terrestrial Eötvös experiment, the fractional difference in oscillation frequency $\Delta f/f$ (where f is the mean oscillation frequency) can be shown to be

$$\frac{\Delta f}{f} \sim \left(\frac{2R}{3|\beta||b|}\right)\frac{\Delta a}{g} \qquad (5.2c)$$

where R is the radius of the orbited body. In the case of a satellite orbiting the Earth ($R = 6 \times 10^6$ m) and a force characterised by the coupling and range parameters specified earlier, the above fractional difference in oscillation frequency is more than *one million* times greater than the fractional change in acceleration.

Testing for new forces by satellite is not without its own problems, and it is unlikely that such an experiment will ever be undertaken soon. For one thing, the differential effect of the Earth on the test mass and shell would be entirely eliminated only if the gravitational field of the Earth were perfectly uniform. Since this is not the case, one must take account of the residual 'tidal' force (the same type of force responsible for the occurrence of ocean tides) resulting from the variation in the strength of the Earth's gravity throughout the interior of the satellite. Secondly, the incentive to find a fifth force has considerably waned, for further examination of the Eötvös paper by other researchers seems to show that the original discrepancies might well have had a far more mundane explanation: air currents in Eötvös's laboratory!

Nevertheless, one never knows when or where something wholly new may crop up. I, for one, would still like to know whether an antineutron falls upward.

* * * * * * * * * * * *

Within the framework of quantum mechanics, the Earth's gravitational field can affect the wave function of an elementary particle in ways for which no classical interpretation in terms of forces can be given. One striking example of this is the effect of gravity on neutrons moving *horizontally*[8]. Imagine a neutron beam incident upon a beam-splitting device that either transmits a neutron or reflects it vertically upward with 50% probability (Figure 5.1). The vertically reflected neutron encounters a perfect mirror that reflects it horizontally so that it propagates a distance L exactly parallel to, but at a height H above, the path followed by a transmitted neutron. The transmitted neutron, after propagating a horizontal distance L, also encounters a perfect mirror that reflects it vertically upward. The two neutron paths, which together form a rectangle, meet at another beam-splitting device that transmits

[8] R. Colella, A. W. Overhauser and S. A. Werner, Observation of Gravitationally Induced Quantum Interference, *Phys. Rev. Lett.* **34** (1975) 1472.

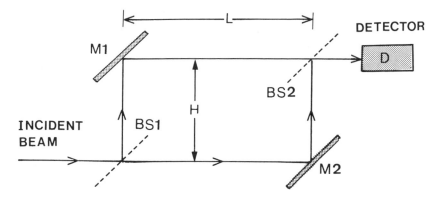

Figure 5.1. Schematic diagram of a neutron Mach–Zehnder interferometer. Beam splitters BS1 and BS2 transmit and reflect a neutron with 50% probability. The plane of the interferometer is vertical so that the path segment of length L between mirror M1 and BS2 is a height H above the corresponding segment between BS1 and mirror M2. The components of a neutron wave transmitted and reflected by BS2 are coherently recombined at the detector D. Only one neutron at a time traverses the apparatus.

an 'upper' neutron and reflects a 'lower' neutron with equal probability (50%) horizontally into a detector. The detected neutrons are counted – but under the circumstances the experimenter cannot know whether a particular neutron has followed the upper or lower horizontal path. The experimental configuration constitutes the neutron counterpart to what in optics is called a Mach–Zehnder interferometer. A classical light wave, however, partitioned at the first beam splitter, traverses both routes to the second beam splitter. It is worth emphasising, therefore, that the neutron flux is ordinarily low enough that only one neutron at a time passes through the interferometer.

According to standard quantum mechanical procedure, to determine the probability of receipt of a neutron at the detector – or, equivalently, the neutron count rate – one must add the probability amplitudes for passage of a neutron along one or the other of the two indistinguishable paths. During the time t that it follows the upper horizontal path, an initially reflected neutron of mass m maintains a gravitational potential energy higher by mgH than that of an initially transmitted neutron that has followed the lower horizontal path. Neutrons following the upper or lower pathways experience no differential effect of the *force* of gravity since both routes include a vertical segment of length H over which gravity does equal work on the particles and a horizontal segment of length L over which no work is done. Nevertheless, the two spatially separated components of the neutron wave acquire a relative phase difference ϕ of the form

$$\phi = 2\pi Ut/h, \tag{5.3a}$$

where $U = mgH$ is the difference in potential energy and h is Planck's constant.

Moving with mean (nonrelativistic) speed v, a neutron has linear momentum mv and covers the distance L in a time $t = L/v$. The corresponding neutron wave function, representable to good approximation by a plane wave, is characterised by a wavelength λ in terms of which the speed may be expressed by means of the de Broglie relation

$$p = mv = h/\lambda. \tag{5.3b}$$

Substitution into relation (5.3a) of the appropriate expressions for U, t and v permits one to write the relative phase in terms of directly accessible experimental quantities

$$\phi = 2\pi m^2 gHL\lambda/h^2 \tag{5.3c}$$

(where it is actually the product of the inertial and gravitational masses that enters expression (5.3c)). The probability amplitude for arrival of a neutron at the detector by two indistinguishable pathways is then of the form

$$\psi \sim 1 + \exp(i\phi) \tag{5.3d}$$

from which it follows that the neutron count rate, proportional to the probability of arrival $P(\phi)$, must depend on the acceleration of gravity g, and vary harmonically with the height difference H, according to

$$P(\phi) = |\psi|^2 = \tfrac{1}{2}[1 + \cos\phi]. \tag{5.3e}$$

(The normalisation factor $\tfrac{1}{2}$ restricts the maximum probability to unity.)

The observation of this neutron interference phenomenon (Figure 5.2) demonstrates convincingly that the Earth's gravity can affect the motion of elementary particles under circumstances where it is not the gravitational force itself, but the difference in gravitational potential energy, that has direct physical significance. Interestingly, it illustrates as well that the equivalence principle may be of questionable validity in the realm of quantum mechanics. As a consequence of the equality of inertial and gravitational masses, a classical object moves through a gravitational field along a mass-independent trajectory. However, the

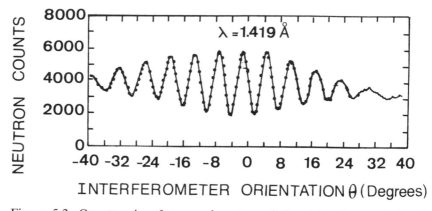

Figure 5.2. Quantum interference of neutrons induced by the gravitational potential of the Earth. The neutron wavelength is approximately 1.42×10^{-8} centimetres. Rotating the interferometer of Figure 5.1 by an angle θ about the incident beam produces a vertical separation $h = H\cos\theta$ between path segments M1–BS2 and BS1–M2. Each experimental point is the result of a total counting time of about seven minutes. (Conceptually inessential differences between the actual interferometer and the idealised interferometer analysed in the text lead to an interference pattern of the form $a + b\cos\phi$, where, in contrast to equation (5.3e), a and b are unequal.) (J.-L. Staudenmann *et al.*, *Phys. Rev.* **A21** (1980) 1419.)

relative phase shift ϕ depends on mass, and the probability of particle arrival, therefore, is not the same for all particles.

＊＊＊＊＊＊＊＊＊＊＊＊

In addition to the force of gravity, which acts whether the Earth turns or not, and the centrifugal force, which any object on the rotating Earth experiences even if stationary relative to the Earth's surface, there is yet another interaction, the Coriolis force, that affects objects *in motion* on the surface of the rotating Earth. The Coriolis force deflects a moving object from apparent straight-line motion, as judged by an observer at rest on the Earth, and, like gravity and the centrifugal force, is independent of all intrinsic chemical and physical properties of an object except that of mass. The resemblance in this way of the Coriolis and centrifugal forces to gravity is illustrative again of Einstein's equivalence principle, the version which asserts that gravity and accelerated motion are locally indistinguishable.

The Coriolis force is another example of a pseudo force in the sense that an observer in an inertial (nonaccelerating) reference frame does not need to invoke it to explain physical events. Imagine two ball players on diametrically opposite ends of a large rotating platform like that of a carousel. One throws a ball towards the other. From the bird's-eye view of a stationary observer above the carousel, the ball moves in a straight line across the surface as the two players rotate with the platform. However, from the perspective of the intended receiver, with respect to whom the thrower has remained motionless, the ball follows a curved path away from the centre as if acted upon by some force – the Coriolis force. Under just the right circumstances, the thrower, himself, can rotate into position to catch the thrown ball. From *his* perspective, the ball has followed a trajectory outward and back again like a yo-yo without a string! In the rotating frame of reference the Coriolis force has physical consequences.

The Coriolis force on an object of mass m moving with speed v along a surface that is rotating about a perpendicular axis with angular velocity ω is proportional to $mv\omega$. The direction of the force depends on the direction of motion of the object and on the sense of rotation of the frame. On the Earth, which spins at an angular rate of 360° in 24 hours, or about 7.3×10^{-5} radians/second, the Coriolis force can markedly affect the patterns of global air flow, although it is ordinarily too weak to influence the local motion of relatively small objects over a time scale short enough that someone would likely have the patience to watch it. Nevertheless, it does have perceptible effects on small objects over sufficiently long intervals of space or time. A directly aimed cannon shot, for example, will in the Northern Hemisphere fall to the right of the target if deflection by the Coriolis force is not taken into account in the design of the sighting mechanism. British sailors rediscovered this fact during a naval engagement with Germany near the Falkland Islands off the south-eastern coast of Argentina early in the First World War. Their sighting mechanisms had been constructed for warfare in the Northern Hemisphere, and consequently their projectiles fell to the left of the German ships by some 100 metres, *twice* the Coriolis deflection.

In a more mundane example, countless visitors to science museums each year are likely to notice a Foucault pendulum. First devised in the mid-nineteenth century by the French physicist, Jean Léon Foucault, the pendulum shaft – sometimes extending several storeys – is suspended vertically above the floor on which is depicted a calibrated ring. As the bob swings back and forth across the ring, the plane of oscillation appears to precess slowly relative to the fixed reference marks. To an inertial observer, it is the floor that rotates under the pendulum at a rate that depends upon the latitude of the site.

That the rotation of the Earth can also affect the motion of an elementary particle was demonstrated in a beautiful experiment again involving the quantum interference of neutrons[9]. As is clear from the effect of gravity on neutron interference or the effect of a confined magnetic field on electron interference (Chapter 1), the concept of energy retains a physical significance under conditions where it would be meaningless to speak of a force. This is the case with the 'neutron Sagnac effect'.

The Sagnac effect, which was first demonstrated with light by the French physicist M. G. Sagnac in 1913, is a phase shift in the interference of two coherent waves as a consequence of the rotation of the interferometer. The geometrical configuration of a Sagnac interferometer resembles that of the Mach–Zehnder interferometer described above, except for one critical detail. The second beam splitter is replaced by a mirror so that the waves reflected and transmitted at the first (and only) beam splitter propagate in opposite directions completely around the interferometer and overlap again at their place of entry. If the interferometer were stationary (or moving at a uniform velocity relative to some other inertial reference frame), the time required for a light wave to complete one circuit about the interferometer would be the same for either direction of propagation. When the interferometer rotates, however, the beam splitter rotates towards one of the waves and away from the counterpropagating wave. Suppose the interferometer is rotating clockwise according to an inertial observer suspended above it. The wave propagating counterclockwise would then complete a circuit in a time interval shorter than that of the clockwise propagating wave. A relative phase difference would therefore develop between the two waves given by $\Delta t/T$, where T is the period (reciprocal of the frequency) of the waves and Δt is the difference in time for the two counterpropagating waves to complete a circuit.

For an interferometer of area A (i.e. the area enclosed by the counterpropagating beams) rotating at angular velocity ω radians/second about an axis inclined at an angle θ to the direction normal to the plane of the interferometer the time difference Δt is given by the approximate expression

$$\Delta t \sim (4A\omega/v^2) \cos\theta, \tag{5.4a}$$

[9] S. A. Werner *et al.*, Effect of the Earth's Rotation on the Quantum Mechanical Phase of the Neutron, *Phys. Rev. Lett.* **42** (1979) 1103; J.-L. Staudenmann *et al.*, Gravity and Inertia in Quantum Mechanics, *Phys. Rev.* **A21** (1980) 1419.

where v is the speed of the wave relative to the nonrotating laboratory. This approximation is good to the extent that one can neglect the square of the ratio of the speed of rotation to the speed of the wave, or $(\omega R/v)^2$, where R is a characteristic size of the interferometer (e.g. the radius, if the light beam followed a circular path). For the case of counterpropagating electromagnetic waves, the speed of propagation is the universal constant c, and the Sagnac phase shift, ϕ_S, expressed in terms of the wavelength $\lambda = cT$ becomes

$$\phi_S = \Delta t/T = (8\pi A\omega/c\lambda)\cos\theta. \qquad (5.4b)$$

Because neutrons have wave-like properties, the rotation of a neutron interferometer should also lead to a phase shift between counterpropagating components of a split neutron beam. In this case, the quantity corresponding to the period of the neutron wave is λ/v, where the speed v is not a universal constant, but is related to the wavelength through relation (5.3b). Substitution of the factors appropriate to a massive particle leads to a corresponding phase shift

$$\phi_S = (8\pi m A\omega/h)\cos\theta \qquad (5.4c)$$

that depends on the (inertial) mass of the particle but is totally independent of velocity and wavelength.

In the neutron Sagnac experiment, the Mach–Zehnder type of interferometer was employed again, but oriented so that the incoming neutron beam and the plane of the interferometer were vertical. From the symmetry of the configuration it should be clear that turning the plane of the interferometer about the vertical axis does not alter the height above ground – and hence the gravitational potential – of any point of the neutron pathways through the interferometer. Such a rotation, therefore, would not change the gravity-induced phase shift. It does, however, reorient the plane of the interferometer (specified by its normal direction) with respect to the rotation axis of the Earth. Thus, the entire variation in the neutron count rate for different settings of the angle θ should be attributable to the Sagnac effect. This intensity variation is expressed by a relation analogous to (5.3e), but with the (now constant) gravitational phase shift augmented by ϕ_S. Actually, since an individual neutron, in going from the first to the second beam splitter (and then to the detector), does not make a complete circuit around the interferometer, but only one-half a circuit, the theoretically predicted Sagnac phase shift should be one-half that of relation (5.4c). The

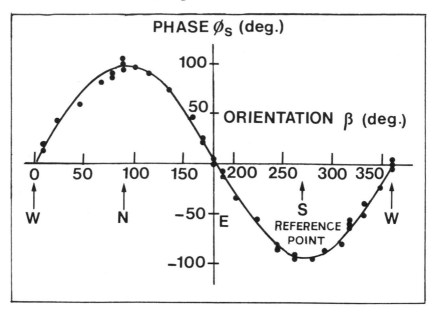

Figure 5.3. Influence of the Earth's rotation on the self-interference of neutrons (neutron Sagnac effect). The interferometer, positioned so that the incident beam is along the vertical direction, is turned about the vertical by an angle β (relative to a reference direction) to vary the angle θ between the normal to the interferometer plane and the rotation axis of the Earth. The graph shows the predicted variation in rotational phase shift ϕ_S as a function of β (full line) and the corresponding experimental points. (The angle θ in equation (5.4c) is related to β by: $\cos\theta = \cos\theta_L \sin\beta$, where θ_L is the latitude of the experimental site.) (Adapted from J.-L. Staudenmann *et al.*, *Phys. Rev.* **A21** (1980) 1419.)

experimental results nicely confirmed this predicted effect of the Earth's rotation (Figure 5.3).

* * * * * * * * * * * *

At the time I began to study the problem of light reflection from an optically active medium (Chapter 4), I also became interested in the effects of the Earth's rotation on quantum mechanical systems. Though outwardly quite different topics, there is an important point of contact that relates them. Both phenomena involve chirally asymmetric interactions.

Consider the rotating Earth. Because the Earth turns towards the east, the Coriolis force on a person running due east along the Equator (or counterclockwise to someone looking down upon the North Pole) is directed radially outward away from the centre of the Earth. If the runner changes direction and proceeds due west along the Equator (clockwise to the observer above the North Pole), the direction of the Coriolis force on him will be radially inward towards the centre of the Earth. The Coriolis force distinguishes between clockwise and counterclockwise – between right-handed and left-handed – senses of motion. To an observer confined to a rotating reference frame like the Earth the Coriolis force is a chirally asymmetric force.

It will be recalled that chiral objects or processes are not superposable on their mirror images. The transformation of an east-bound runner into a west-bound runner is effected by reflection of the Earth and runner in a mirror. In such a reflection, both the direction of the runner *and* the sense of rotation of the Earth are reversed in which case the Coriolis force continues to point outward. However, an Earth that turns towards the west simply does not exist – although presumably there is no physical law forbidding such a historical possibility. In any event, the original scene and its mirror image are not superposable.

Although the neutron is believed to be composed of three basic particles (quarks), the internal structure of the neutron has nevertheless played no significant role in the self-interference experiments sensitive to the gravity and rotational motion of the Earth. These experiments prompted me to wonder, however, about the quantum effects of gravity and rotation on the internal dynamics of composite quantum systems like atoms and molecules. Because the Coriolis force distinguishes left- and right-handed senses of motion, could it by any chance give rise to optical activity in atoms? Would such an effect be observable?

Discussions of the physics of atoms almost always take for granted at the outset that the frame of reference is not accelerating. The laws of quantum mechanics were initially formulated for inertial frames, and actual experiments on atoms are ordinarily executed under such conditions that this assumption would appear adequate. The Earth is not, of course, a true inertial reference frame. However, the Coriolis force of the Earth's rotation on an atomic electron is smaller than the electrostatic force binding the electron to the nucleus by a factor of about one hundred billion billion (10^{20}). This is very small, indeed! (The centrifugal force of the Earth on a bound electron is at least four orders of magnitude smaller.) To detect an influence of the Coriolis force of the Earth in the optical properties of an atom would be tantamount to observing one of the weakest interactions by far in which an atom has participated[10]. And yet this prospect may not be entirely hopeless.

[10] There are, of course, still weaker interactions, as for example that of an atom with 'gravity waves', the rippling of the space-time continuum itself, produced by (among other means) catastrophic collapse of some massive astrophysical object.

It is interesting to speculate that, weak though it may be, the effect of the Earth's rotation on individual atoms could conceivably be connected with one of the outstanding unsolved problems in the life sciences: the origin of biomolecular chirality. Why living things make and use molecules of specified handedness such as right-handed sugar molecules or left-handed amino acids, is not known. Perhaps, over the aeons during which the molecules of life evolved, the chirally asymmetric effect of the Earth's rotation may have led to a preferential molecular handedness in one hemisphere that, through the random accidents of history, spread over the entire planet.

Specifically, how can the rotation of the Earth influence the structure of an atom? Although classical mechanics does not in general serve as an adequate basis for understanding the dynamics of an atom, there are times when the imagery of classical physics provides insight, at least when the quantum mechanical degrees of freedom involved have classical counterparts. It is worth stressing at the outset that it is the *internal* dynamics in which we are interested here (both classical and quantum mechanical) – i.e. in the motion of the electrons relative to the nucleus, and not the motion of the 'centre of mass' of the atom. The centre of mass of a system of particles – which need not correspond to the location of any actual particle – moves in response to the net *external* force as if all the mass of the system were concentrated at that hypothetical point. The rest frame of the system is the reference frame in which the centre of mass is stationary. For ordinary atoms (in contrast to exotic atoms) in which all bound particles are electrons, the centre of mass coincides with the location of the nucleus to good approximation.

Consider for simplicity a planetary atom with a single electron in circular orbit about the nucleus at an angular frequency ω_0 radians/second (as determined theoretically for an atom at rest in an inertial frame). In fact, to take the simplest case possible, locate the atom at the North Pole so that the axis about which the electron revolves coincides with the rotation axis of the Earth. To an inertial observer suspended above the North Pole, the angular frequency of the electron is ω_0 irrespective of the sense (clockwise or counterclockwise) of the revolution. However, to an observer fixed on the Earth which turns, let us say, at ω radians/second, the angular frequency of the electron is $\omega_0 - \omega$ if the electron revolves in the same sense as the Earth rotates, and $\omega_0 + \omega$ if the electron revolves in the opposite sense. Even though an observer cannot actually 'see' the motion of an electron in an atom, he would nevertheless draw the preceding conclusions by correlating the frequency and circular polarisation of the spontaneously emitted radiation. A circulating charged particle is undergoing periodic acceleration and, according to classical electrodynamics, emits along the rotation axis electromagnetic waves with transverse electric fields that rotate in the same sense and at the same frequency as the orbital motion of the electron. A real atom, of course, does not continuously radiate – or it would collapse practically instantaneously. We will see, however, that quantum mechanics sustains the foregoing picture of chirally inequivalent orbital motions.

In addition to the characteristic spontaneous emission of radiation, the optical response of an atom to incident radiation can also be influenced by the rotation of the Earth. The index of refraction of a material, as I discussed in the previous chapter, was shown by Maxwell to be effectively equal to the square root of the dielectric constant in the case (relevant to the present discussion) that the material is not intrinsically magnetic. The dielectric constant is, itself, a measure of the extent to which the bound electrons of the sample are displaced from their equilibrium positions by an external electric field, such as the electric field of an incident light wave. The greater the displacement, the greater will be the electric dipole moment of an individual atom (which is the displacement multiplied by the electron charge), the greater will be the resulting dielectric constant (which grows with the number of electric dipoles in the sample), and hence the greater will be the corresponding refractive index of the material.

It is the atomic polarisability α that expresses the proportionality between the displacement of a bound electron from its equilibrium position and the strength of the applied electric field. To determine the polarisability (still within the framework of classical mechanics) one solves Newton's equations of motion for the *forced* motion of the electron at the frequency of the incident light wave. This is a standard and relatively elementary problem for an atom in an inertial reference frame. The electron is then subject to the electrostatic binding force, possibly some damping force that takes account of energy loss by spontaneous emission, and the driving force of the electric field of the light. The magnetic field of the light wave can usually be neglected, for it results in a force weaker than that of the electric field by the ratio of the electron speed to the speed of light. Disregarding the effects of damping and magnetism, and assuming an incident light wave of angular frequency Ω, one arrives at the following simple expression for the atomic polarisability

$$\alpha(\Omega) = (e^2/m)/[\omega_0{}^2 - \Omega^2], \qquad\qquad (5.5a)$$

where e is the electron charge and m is the electron mass. Note that the polarisability increases as the frequency Ω of the light approaches the 'resonance' frequency ω_0 of the atom (or molecule). Thus the index of refraction of a transparent material like glass, for which the resonance frequencies typically fall in the ultraviolet portion of the spectrum ($\omega_0/2\pi \sim 10^{15}$ Hz), is larger for blue light ($\Omega/2\pi \sim 6.3 \times 10^{14}$ Hz) than for red light ($\Omega/2\pi \sim 4.4 \times 10^{14}$ Hz); blue light will correspondingly be refracted to a greater extent than red light as it enters the glass from air. To a good approximation the index of refraction (n) of a sufficiently rarefied sample of atoms that behave independently of one another is related to the atomic polarisability in the following way

$$n \sim 1 + 2\pi N\alpha, \qquad\qquad (5.5b)$$

where N is the number of atoms per unit of volume.

Return now to the classical atom at the North Pole of the rotating Earth (Figure 5.4). As analysed by an Earth-bound observer, the orbiting electron is subject, not only to the forces described above, but also to the Coriolis and centrifugal forces. Also, to repeat, it is not the characteristic motion of the electron that is of concern now, but only the motion engendered by the electric field of the incident light wave. Imagine a light wave of frequency Ω, as measured by the *Earth-bound observer*, propagating upward along the common rotation axis of the Earth and electron. If the wave is left circularly polarised it drives the electron about its centre of attraction, the atomic nucleus, in the same sense as the Earth rotates; a right circularly polarised wave drives the electron in the opposite sense. If the electron revolves in response to a left circularly polarised wave, the Coriolis force accelerates it radially outward (i.e. outward along the radial line from the nucleus to the electron, not from the centre of the Earth to the electron), thereby increasing the displacement of negative and positive charges within the atom. This leads to a larger electric dipole moment. Conversely, the Coriolis force accelerates an electron moving in response to a right circularly polarised wave radially inward, leading to a smaller electric dipole moment. The centrifugal force on the electron is directed radially outward irrespective of the sense of circulation.

In sum, from the perspective of an Earth-bound observer, the sample of atoms exhibits a larger index of refraction for left circularly polarised light than for right circularly polarised light. This is exactly what is

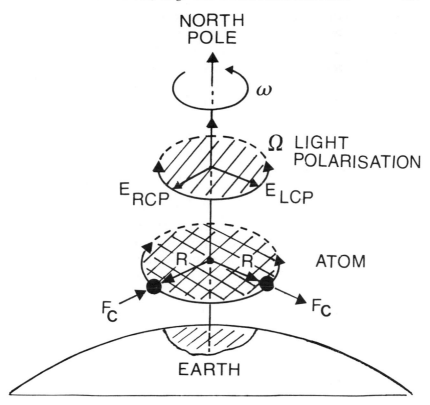

Figure 5.4. Heuristic model of the effect of the Earth's spin on the dynamics of a classical atom in the simple case where the Earth and bound electron rotate about a common axis. When driven by the electric field (E) of an incident left circularly polarised (LCP) light wave of angular frequency Ω, the electron orbits the nucleus in the same sense as the Earth spins; the Coriolis force (F_C) on the electron acts radially outward. When driven by an incident right circularly polarised (RCP) wave of the same frequency, the electron experiences a Coriolis force radially inward. To an inertial observer above the Earth there is no Coriolis force, but the frequencies of the two light beams are respectively $\Omega + \omega$ and $\Omega - \omega$, where ω is the spin angular frequency of the Earth.

required for the atoms to be optically active. The existence of an atomic circular birefringence (difference in chiral refractive indices, $n_L - n_R$) has been inferred for the special orientation of an atom at the North Pole, but the conclusion holds generally for any location on the Earth although the strength of the predicted effect varies with the relative orientation of the light beam and the Earth's axis.

I have noted previously that the Coriolis force is termed a fictitious force originating in the acceleration of the reference frame. Does this mean that the predicted optical activity is, itself, fictitious? Would the inertial observer suspended above the North Pole agree that the atoms exhibit a chiral asymmetry? To answer the question let us examine the expressions derived by the Earth-bound observer for the polarisability of a rotating atom. The two expressions for left and right circular polarisations are similar in form to that of relation (5.5a) derived for an atom in an inertial reference frame

$$a_L = (e^2/m)/[\omega_0^2 - (\Omega + \omega)^2], \tag{5.6a}$$

$$a_R = (e^2/m)/[\omega_0^2 - (\Omega - \omega)^2]. \tag{5.6b}$$

To the inertial observer, however, the above relations are in fact the same relation as (5.5a) – only evaluated for different frequencies. If the frequency of the left circularly polarised light wave propagating upward along the rotation axis is Ω relative to the Earth-bound observer (who is himself rotating in the same sense at the frequency ω of the Earth), then the inertial observer would declare the light frequency to be $\Omega + \omega$. Similarly, the right circularly polarised wave, also of frequency Ω to the Earth-bound observer, would present a frequency $\Omega - \omega$ to the inertial observer.

The inertial observer might therefore say: 'Of course the index of refraction is different for the left and right circularly polarised waves. Their frequencies are different, and it is well known that a higher frequency leads to a larger refractive index. The atoms, however, are chirally symmetric'. To this the Earth-bound observer could in truth reply: The frequency of both waves is the same. The Coriolis force produces chirally asymmetric atomic polarisabilities'. Both interpretations are correct. Still, it is useful to keep in mind that, as denizens of a rotating reference frame, physicists ordinarily interpret the results of their measurements in terms of the apparatus and interactions in their own stationary Earth-bound laboratories, and do not feel constrained to consult inertial colleagues suspended above the planet.

Since the internal dynamics of actual atoms are not accurately described in terms of electron trajectories influenced by Newtonian forces, one might wonder whether the foregoing classical analysis is in any way reliable. In brief, the answer is basically affirmative – with one important *caveat*. It is understood, of course, that where an objectively real physical quantity like an orbital radius might appear in the mathematical expressions of classical mechanics, the analogous quantum mechanical expressions would contain matrix elements (i.e. integrals) of a corresponding operator, providing in effect a measure of the likelihood that the atom can undergo certain transitions between its states. If the matrix elements connecting particular quantum states of interest vanish, then quantum mechanics does not permit the designated process to occur even though classical mechanics may have yielded a seemingly respectable non-null result. I shall give an important example of this shortly.

Within the framework of quantum mechanics, the interactions that affect the internal state of an atom are incorporated in the appropriate equation of motion (e.g. the Schrödinger equation) not as forces, but as contributions to potential energy. For an atom rotating with the Earth, the effects on its constituents of both the centrifugal and Coriolis forces may be shown *classically* to derive from an extra energy term, (U_R), that involves a 'coupling' of the internal angular momentum (L) of the atom to the angular velocity (ω) of the Earth as follows

$$U_R = -\omega L \cos\theta, \tag{5.7a}$$

where θ is the angle that the atomic angular momentum makes with the rotation axis of the Earth. Again, by 'internal' angular momentum I mean the orbital motion of the electron about the atomic nucleus, and not the movement of the whole atom about the axis of the Earth. Classically, the angular momentum of an object of mass m moving about a centre of attraction with speed v in an orbit of radius R has the magnitude mvR and is oriented perpendicular to the plane of the orbit in a 'right-hand sense'. That is, if one wraps the fingers of his right hand about the orbit so that they point in the direction in which the object circulates, then the extended thumb gives the direction of the angular momentum.

In stark contrast to the classical picture of an atom, however, the details of electron motion within a quantum mechanical atom cannot be pictured. How, then, is the electron angular momentum to be oriented? In fact, quantum mechanics does not permit one to know this orientation. One can know only the magnitude of the angular momentum, which is restricted to integer multiples of $\hbar$, and the projection of the angular momentum along an arbitrary axis. In the present case it is convenient to choose this axis to be the axis of the Earth. For an electron angular momentum of magnitude $L = \sqrt{\ell(\ell+1)}\,\hbar$, the projection $L\cos\theta$ must then take values $M\hbar$ in which the azimuthal or magnetic quantum number M spans the range of $2\ell + 1$ integers from $-\ell$ to ℓ in steps of 1. Projections that differ only in sign refer to electron states that differ only in the sense of electron circulation about the quantisation axis.

If an atom in an inertial reference frame is not subject to external perturbations, then all directions of the quantisation axis are equivalent, as are also the two senses of rotation about the axis. One would therefore not expect the energy of a quantum state to depend on the orientation of the quantisation axis or on the azimuthal quantum number. On the spinning Earth, however, matters are different, for now there is a singular sense of motion about a particular direction.

Substitution of the potential energy (5.7a) into the Schrödinger equation to determine the energy eigenstates of the electron from the perspective of an Earth-bound observer yields the following interesting result. The state energy, E, expressible in the form

$$E = E_0 - M\hbar\omega \qquad (5.8)$$

(where E_0 is the corresponding energy on a nonrotating Earth), now depends on the component of the electron's orbital angular momentum along the rotation axis of the Earth. If the electron circulates in the same sense as the Earth rotates (i.e. M is positive), the energy of the quantum state is lowered by $M\hbar\omega$. Conversely, the energy is raised by $|M|\hbar\omega$ when the electron and the Earth rotate in opposite senses, and M is negative. This result is in complete analogy to the previous classical treatment leading to orbital frequencies $\omega_0 \pm \omega$ and, again, may be tested experimentally by examination of spontaneous emission. A quantum mechanical atom radiates when the bound electron undergoes a 'quantum jump' to a lower energy state. If an electron in a quantum state of nonzero angular momentum undergoes an electric dipole transition to a lower state of zero angular momentum, the emitted photons will have two possible frequencies: $\omega_0 \pm \omega$, where ω_0 is the corresponding frequency in an inertial reference frame.

The quantum mechanical analysis of a rotating atom interacting with incident left and right circularly polarised light involves the use of a mathematical procedure termed perturbation theory which will not be described here. I note only that the calculation justifies the classical picture of the inequivalent action of the Coriolis force on counter-circulating electron orbits to create chirally asymmetric polarisabilities. The end result is the prediction of a very weak circular birefringence $n_L - n_R$ that can be on the order of about 10^{-18} for light falling in the visible and ultraviolet regions of the spectrum. A circular birefringence of this magnitude would lead to the minute optical rotatory power of about 10^{-11} degrees per metre of material! As stated before, this

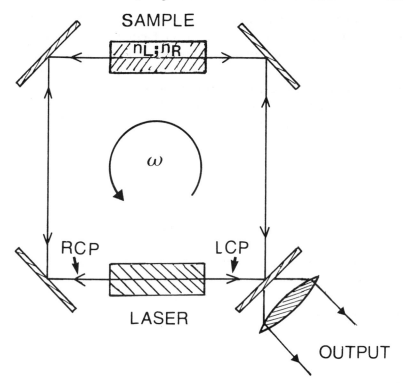

Figure 5.5. Schematic diagram of a rotating ring laser interferometer with a sample that displays circular birefringence (different refractive indices n_L and n_R for the two forms of circularly polarised light). RCP and LCP light waves propagate in opposite senses about the ring and give rise to a beat frequency produced in part by the interferometer rotation and in part by the circular birefringence.

optical activity is far smaller than already observed in atoms as a consequence of the weak nuclear interactions.

Is so weak an optical effect detectable? The answer, a guarded affirmative, brings us back to the Cashmere Cavern and the Canterbury ring laser. In a ring laser, as in the Sagnac interferometer, two coherently produced light waves traverse the same closed path in opposite directions (Figure 5.5). What distinguishes the laser from a passive (i.e. no gain) interferometer, however, is the presence within the ring of a medium with a population inversion such as discussed in Chapter 4. Although the laser may emit light over a range of frequencies, only those oscillations are sustained which satisfy a resonance condition whereby an integral number of wavelengths span the perimeter of the ring. Those waves that do not satisfy this requirement are effectively suppressed by destructive interference. In fact, all waves (light, sound, water, etc.) in a closed container or cavity are subject to the imposition of boundary conditions.

When the ring laser is rotating, the effective length of the trip around the cavity – and hence the resonant wavelength and frequency – is different for the clockwise and counterclockwise modes. Upon recombination at a detector, the two modes, no longer synchronised, produce a beat frequency similar in principle to the beat heard when one strikes two neighbouring keys on a piano. There is an advantage to measuring a beat frequency in a ring laser compared with a phase shift in a passive interferometer. For one thing, frequency is an experimental quantity that can be measured with relative facility and to very high precision. (As one example, the frequency corresponding to the hyperfine splitting in ground state hydrogen can be measured to better than one part in 10^{12}.) For another, the factor relating the beat frequency to the optical path length difference is larger (by the ratio of the speed of light to the circumference of the ring) than the corresponding factor relating the Sagnac phase shift to the optical path length difference.

In addition to providing a highly sensitive monitor of rotation, a ring laser permits one to measure small optical anisotropies, such as the birefringence of a material. Suppose that the natural modes of the ring laser are circularly polarised, and that the laser could be excited bidirectionally with left and right circularly polarised waves traversing the ring in opposite directions. The presence of a sample of optically active matter in the ring would likewise give rise to a frequency difference, or beat frequency, since the left and right circularly polarised waves are retarded by the sample to different extents. Clearly, the case of two counterpropagating waves retarded unequally with respect to a stationary laser is equivalent in principle to that of a moving laser 'gaining' on one wave and 'receding' from the other. This frequency shift is linearly proportional to the circular birefringence and the mean operating frequency of the laser.

For the problem at hand, it is the rotation of the Earth – and hence of the ring laser fixed to the Earth – that *generates* the optical activity in the sample. Since (as in the case of the Eötvös experiment) one cannot stop the Earth from rotating, how will the experimenter know that a shift in beat frequency has occurred? The solution to the problem is effectively the same as that employed by Eötvös: make a comparative measurement. Although the Earth's rotation should induce optical activity in all terrestrial matter, the magnitude of this circular birefringence depends on both the quantum level structure of the material and the frequency of the light.

Circular birefringence is a nonresonant phenomenon; that is, the material is essentially transparent to the light employed. Nevertheless, the circular birefringence of a material, like the atomic polarisability, ordinarily increases as the frequency of the light approaches, but never falls within, the region of the spectrum where absorption occurs. Introduction into the ring, therefore, of a sample material for which the laser frequency is close to an atomic transition should shift the beat frequency attributable to rotation, alone, by a small but significant amount. Subsequent reversal of the direction of the counterpropagating left and right circularly polarised waves – in analogy to the exchange of masses in Eötvös' torsion balance – can be shown to shift the beat frequency in the opposite direction. Unlike the Eötvös experiment, however, a null result is not expected.

Theoretical analysis of the ideal performance of a ring laser[11] suggests that a laser of the type and size (area of 1 square metre) now being developed at the University of Canterbury should be able to detect a shift in frequency smaller than the operating frequency itself by a factor of about 10^{19}. If this ideal performance is actually realisable, the predicted atomic circular birefringence would just marginally fall within the resolution capacity of the laser.

If the optical activity induced by the Earth's rotation is far weaker than the atomic optical activity resulting from nuclear interactions, how is it to be distinguished from the latter? Actually, the two types of optical activity differ significantly in their symmetry properties. Because weak neutral current interactions between the nucleus and orbiting electrons actually mix close-lying atomic states of opposite parity (for example, the S and P states within a given electronic manifold), the resulting optical activity is similar in nature, although different in magnitude, to the molecular optical activity (discussed in Chapter 4) associated with chirally asymmetric three-dimensional chemical structures. In this type of optical activity the sense of chiral asymmetry is defined with respect to the direction of propagation of the light beam, there being no other preferred direction in an optically isotropic material. One consequence of this, pointed out earlier, is that a plane-

[11] G. E. Stedman and H. R. Bilger, Could a Ring Laser Reveal the QED Anomaly via Vacuum Chirality?, *Phys. Lett.* A122 (1987) 289.

polarised light beam reflected back upon itself to its point of entry in the optically active medium shows no net optical rotation.

The optical activity generated in atoms by the Earth's rotation is different – and illustrates by that difference an illuminating connection between rotation and magnetism. There is a well-known theorem in classical mechanics known as Larmor's theorem (derived in 1897 by J. J. Larmor, who in that same year also derived the Larmor formula employed in Chapter 2) which states that the effect of a constant magnetic field B on a system of particles of mass m and charge q is to superimpose on its normal motion (i.e. in absence of the magnetic field) a uniform precession. The angular velocity of precession, the magnitude of which is designated the Larmor frequency, may be shown to be

$$\omega_L = - qB/2mc. \tag{5.9a}$$

The theorem is not exact, but rather an approximation valid to the extent that one can neglect terms of order B^2 and higher. Because of the negative sign in (5.9a), the orientation of the angular velocity is *opposite* the orientation of the magnetic field if the particles are positively charged. Another way of expressing this is to say that the precession occurs in the same sense as the current of negative electrons that generates the magnetic field in the first place. In any event, the basic idea is that one can sometimes simplify the analysis of a system of particles in a magnetic field by eliminating the field and placing oneself in a frame of reference rotating at the Larmor frequency.

The influence of the rotation of the Earth on atoms can be regarded in some ways as the *converse* of Larmor's theorem. In other words, the effect of a uniform global rotation at angular velocity ω on a system of particles in an environment free of electric and magnetic fields is equivalent to that of a constant magnetic field (which, for consistency, I shall call the Larmor field B_L) obtained by rearranging relation (5.9a)

$$B_L = - 2mc\omega/q. \tag{5.9b}$$

Thus, the rotational separation of magnetic substates degenerate in an inertial frame (relation 5.8)) is the analogue of the Zeeman effect, the splitting of degenerate states by a static magnetic field. In fact, the potential energy term (5.7a), with ω replaced by the equivalent expression from equation (5.9b), has exactly the form of an interaction of a magnetic field with a magnetic dipole moment of magnitude $qL/2mc$. (This is the magnetic dipole moment one would expect for a current loop consisting of a single charged particle in circular orbit with angular momentum L.) Likewise, the phase shifts produced in the Sagnac effect by rotation and in the Aharonov–Bohm effect by a magnetic flux are analogous – the connection being particularly direct when expressed in terms of the vector potentials from which the corresponding magnetic fields derive. In a similar way, rotational optical activity has a magnetic analogue, the Faraday effect, discovered by Michael Faraday in 1845. The plane of polarisation of a linearly polarised light beam transmitted through an isotropic dielectric in a static magnetic field is rotated by an amount linearly proportional to the magnetic field strength and the path length through the medium.

From a classical perspective the Faraday effect is produced by the inequivalent action of the Lorentz force on oppositely circulating electron orbits. Replacing the Lorentz force by the Coriolis force, or equivalently, substituting the Larmor field B_L for the actual magnetic field appearing in the calculation of the Faraday rotation, yields the classical mechanical expression for rotational optical activity. The optical rotation, however, of the Faraday effect occurs with respect to the fixed magnetic field, and not with respect to the light propagation direction. The practical consequence of this is that the net optical rotation of a light beam reflected back upon itself to its point of entry into the sample is *twice* that of a single passage – not zero as in the case of structural optical activity. This same symmetry holds for optical activity engendered by the Earth's rotation, for which the rotation axis of the Earth replaces the magnetic field direction, and helps serve to distinguish the predicted effect from optical manifestations of the weak nuclear interactions.

The weak nuclear interactions are not only weak, but of extremely short range. I have mentioned previously that, according to the electro-weak theory, the nucleons and bound electrons of an atom are coupled by the exchange of a Z^0 boson, a particle whose mass is about 100 times the proton mass. In quantum mechanics the range of a force can often be estimated quickly by means of the uncertainty principle. The linear momentum of an exchanged (or virtual) particle of mass m can span a range of values from 0 to about mc. Hence the uncertainty, in its spatial location, is approximately

$$\lambda_C = h/mc, \tag{5.10}$$

the so-called Compton wavelength. The Compton wavelength of the Z^0 boson is about 10^{-15} cm, some seven orders of magnitude smaller than the Bohr radius which sets the scale of atomic size. Consequently, weak neutral currents can directly influence only those atomic states – the S states – for which there is significant overlap of electron and nuclear wave functions. For all states of nonzero orbital angular momentum, the electron wave function has a 'node' or zero amplitude at the nucleus (treated in first approximation as a point mass).

In the electronic manifolds of principal quantum number $n = 2$ and higher, the weak nuclear interactions mix close-lying S and P states thereby giving rise to an atomic optical activity that increases in strength approximately with the cube of the atomic number. Therefore, by employing light atoms, using light of such frequency as to avoid contributions from electrons in S states, and by taking advantage of the cumulative enhancement of multiple passes through the sample, one might hope to observe the chiral effects of the Earth's rotation in the domain of atomic physics.

* * * * * * * * * * * *

Although a heuristic explanation of the interaction between an atom and the spinning Earth in terms of the classical Coriolis force seems to have led so far to results in basic accord with those of quantum mechanics, this need not always be the case. Quantum mechanics embraces degrees of freedom for which there are no classical counterparts. Consider, for example, an unexcited hydrogen atom. Based on the classically derived potential energy expression (5.7a), the internal dynamics of an atom in a 1S ground state would be entirely unaffected by the rotation of the Earth since a 1S electron has zero orbital angular momentum. (The centre of mass of the hydrogen atom would still, of course, be subject to the Coriolis force of the Earth's rotation.) This expectation is not correct, however.

An elementary particle can have a nonclassical degree of freedom, spin, which also contributes to its angular momentum. The electron and the proton, for example, are both spin-$\frac{1}{2}$ particles. Although one can try to picture the spin of an electron as analogous to the diurnal rotation of the Earth about its axis, this is not really satisfactory. High-energy scattering experiments probing the internal structure of the electron indicate (in contrast to the proton and neutron) that the electron is a 'point' particle composed of no more fundamental subunits to within an experimental limit of about 10^{-16} cm. If one models the electron as a spinning charged sphere of radius equal to the so-called classical electron radius, $r_0 = e^2/mc^2 \sim 10^{-13}$ cm – deduced by equating the electron rest mass energy mc^2 to its electrostatic potential energy e^2/r_0 –

the resulting linear velocity of a point on the 'equator' of the electron surface would be[12]

$$v = 1.25c/\alpha_{\text{fs}} \tag{5.11}$$

(with α_{fs} again the fine structure constant $e^2/\hbar c \sim \frac{1}{137}$) which exceeds the velocity of light by a factor of over 170. If a smaller radius is adopted, then the violation of relativity is even greater. No classical model of electron structure, in fact, has proved adequate. It seems, therefore, that spin must simply be accepted, and not structurally interpreted.

With account taken of the nonclassical attribute of spin, a completely quantum mechanical analysis replaces relation (5.7a) with the potential energy expression

$$U_R = -\boldsymbol{\omega}\cdot\boldsymbol{F} \tag{5.7b}$$

involving the projection onto the quantisation axis of the *total* internal angular momentum (F) of the atom comprising a vector sum of the electron orbital angular momentum (L) and angular momenta contributed by electron spin (S) and nuclear spin (I).

In the hydrogen ground level the total angular momentum derives exclusively from the spins of the electron and proton which may be oriented either parallel or antiparallel to one another. As a result of the magnetic coupling of electron and nuclear spins, the $n = 1$ level of the hydrogen atom is composed of four hyperfine states. There is a single ground state of zero total angular momentum which, in the notation of Chapter 2, can be designated $1S_{\frac{1}{2}}(F = 0; M = 0)$. In this state the electron and proton spins are oppositely directed. Above the ground state lie the three states $1S_{\frac{1}{2}}(F = 1; M = +1)$, $1S_{\frac{1}{2}}(F = 1; M = 0)$ and $1S_{\frac{1}{2}}(F = 1; M = -1)$ of total angular momentum $1\hbar$ that result when the electron and proton spins are parallel. In an inertial reference frame these three states of total quantum number $F = 1$ are degenerate. If the quantum mechanical treatment leading to relation (5.7b) is correct, the rotation of the Earth should split the energy of the two states with azimuthal quantum numbers $M = \pm 1$ by an amount $\hbar\omega \sim 5 \times 10^{-20}$ electron volts in accordance with relation (5.8). The hyperfine level splitting E_0 between the two rotationally unaffected 1S states with $M = 0$ corresponds to a microwave photon of frequency 1420 MHz and

---

[12] The angular momentum of a sphere of radius r and mass M rotating at angular frequency ω about an axis through the centre is $(\frac{2}{5})Mr^2\omega$. Replacing r with r_0 above and setting the angular momentum equal to $\frac{1}{2}\hbar$ leads to an equatorial linear velocity $v = \omega r = 5\hbar/4Mr_0$ which reduces to relation (5.11).

wavelength 21 centimetres. (This radiation is of much interest to radio astronomers and astrophysicists who search the skies for, among other things, interstellar clouds of atomic hydrogen gas.)

Despite the fact that, according to relation (5.7b), an atomic 1S state ought to be affected by the spin of the Earth, it does not follow from what has been said so far that atoms in the 1S state must necessarily exhibit optical activity when illuminated with microwave radiation. In fact, at first thought, it might appear that such optical activity cannot occur. All the states of the $n = 1$ ground level have the same (even) parity, i.e. the same behaviour under reflection; symmetry rules strictly forbid electric dipole transitions between states of the same parity. The resulting atomic polarisability – and therefore the dielectric constant of a bulk sample of atoms – would not be differentially affected by left and right circularly polarised microwaves.

Nevertheless, quantum theory does predict a rotational optical activity near 1420 MHz. This is one of the occasions where the magnetic field of the incident radiation cannot be neglected. The electron, although a 'point' particle, has not only an electric charge, but also an intrinsic magnetic moment. According to classical electromagnetism an orbiting charged particle constitutes a simple current loop with a magnetic dipole moment. Although no such loop can be envisioned for a spinning electron with zero orbital angular momentum, there is still an intrinsic magnetic moment deriving from (and proportional to) the electron spin angular momentum. To an observer on the rotating Earth the electron magnetic moment would appear to precess about the Earth's rotation axis – a result in keeping with the previously expressed analogy between rotation and magnetism[13].

The interaction of this precessing magnetic moment with the magnetic field component of an electromagnetic wave ultimately gives rise to a magnetic permeability μ that is different for left- and right-circular polarisations. Since the index of refraction depends not on the dielectric constant (ε) alone, but on the product $\varepsilon\mu$, a linearly polarised microwave beam passing through the sample of unexcited Earth-bound hydrogen atoms should undergo optical rotation. Surprisingly, under appropriate circumstances, this optical rotation can be several orders of magnitude larger than that attributable to the orbital motion of the electron.

[13] The magnetic moment of a current loop in a uniform magnetic field experiences a torque produced by the Lorentz force as a result of which it precesses like a gyroscope about the field direction. Although not describable as a current loop, the magnetic moment arising from particle spin undergoes a similar precession when acted upon by a magnetic field.

All things being equal – resonance frequency, radiation frequency, number of atoms per unit volume, etc.– magnetic interactions are ordinarily weaker than comparable electric interactions by the square of the fine structure constant, $\alpha_{fs}^2 \sim 5 \times 10^{-5}$. How, then, can the rotational optical activity associated with electron spin exceed that associated with electron orbital motion? The answer is that all things here are not equal; in particular, the frequencies of (virtual) electric dipole transitions that contribute to rotational optical activity in the visible and ultraviolet are some five or six orders of magnitude larger than the 1S hyperfine transition frequency. The significance of this is as follows.

Recall that, as long as the atoms do not absorb the incident light, the birefringence of a sample increases as the light frequency approaches a resonance frequency. The frequency at which absorption can occur, however, is not infinitely sharp. First, the energy levels themselves have a natural width resulting from their finite lifetime (from spontaneous emission). Second, if the atoms are moving about randomly as a result of thermal motion, the absorbed light will extend over a range of Doppler-shifted frequencies. And third, if the sample is sufficiently dense, the atoms will collide with one another thereby increasing the energy uncertainty of the states.

At low density it is the Doppler effect that principally determines the range of frequencies over which absorption occurs; the extent of Doppler broadening, it is important to note, is proportional to the resonance frequency. It is the potentially very narrow Doppler width of the hydrogen hyperfine transition that allows one in principle to probe the atom with microwave frequencies lying much closer to a resonance than would be possible with visible and ultraviolet radiation. Unfortunately, the resulting optical activity is still extremely weak and, as the technology for state-of-the-art polarimetry is far more advanced for high-frequency electromagnetic waves than for microwaves, the calculated enhancement is unlikely to be of much experimental help at the present.

Any attempt to measure the optical activity of atoms induced by the Earth's rotation will have to overcome some formidable experimental hurdles. One of the most daunting is that of the Earth's own magnetic field which, with a strength of approximately one-half gauss, gives rise to a true Faraday rotation larger than the sought-for effect by some eleven orders of magnitude[14]. However, the field of quantum magnetometry – the measurement of ultra-small magnetic fields by means of superconducting quantum devices (SQUIDs) – has already

---

[14] From equation (5.9b) the 'Larmor field' corresponding to the Earth's angular frequency of rotation (7.3×10^{-5} radians/second) is 8.3×10^{-12} gauss.

achieved wonders; fields as weak as 10^{-11} gauss can be measured. The question is whether an extant field can be shielded to that low value.

The pursuit of an interaction between the rotating Earth and an Earth-bound atom raises another, perhaps more basic and thought-provoking, question as well. What, in fact, does it *mean* to say that an atom rotates? Mathematically, quantum mechanics provides a formal procedure for expressing any wave function (or operator) in terms of the coordinates of a rotated reference system; this is termed the passive view of rotation. The active view, whereby the wave function itself is rotated with respect to a fixed frame of reference, is considered – again mathematically – to be entirely equivalent. It was by application of such transformations that the equations of motion of a rotating quantum system have been derived, and the attendant phenomenon of rotational optical activity inferred.

But does the rotational displacement of an atom – which must necessitate in some way the physical coupling of the atom to its environment through forces of constraint – actually imply as well that the bound electrons are so coupled? This is not an idle enquiry arising from the paradox-laden terminology of classical physics. The issue can be settled in the laboratory: if the frame of reference rotates, but the atom does not, there will be no rotational optical activity. Experiments to search for chiral asymmetries in the interactions of atomic gases or vapours rapidly spun on a laboratory turntable are now under development. At the rates of mechanical rotation achievable (about 100 revolutions per second), the expected optical rotary power (or companion phenomenon of circular dichroism), if it exists, should be measurable.

For nearly two centuries optical activity has been a catalyst in the progress of science both through the research undertaken to understand it, and as an experimental tool to investigate other phenomena. It does not appear even now to be an exhausted subject.

6

The Wirbelrohr's roar
(. . . or rather whistle)

With all due respect to Robert Boyle, there is a 'spring' to the air that that venerable English physicist never dreamed of some three centuries ago when he introduced his fellow natural philosophers to the effects of pressure[1]. Air is not merely compressible; it can course and caper through appropriate devices in such ways as to please the ear and titillate, if not confound, the intellect. I learned that first hand from playing.

Most people I have encountered, for whom physics is anything but relaxation, could hardly imagine 'physics' and 'play' in the same sentence – except, perhaps, one denying their equivalence. Yet the same laws that govern the erudite matters to which physicists give their attention also apply to recreation. Indeed, sometimes nature's subtlest wiles may be invested in the simplest child's toy. One of my favourite photographs[2] shows Wolfgang Pauli and Niels Bohr hunched over the ground observing the behaviour of a Tippetop, a curious little object that, shortly after being spun on its wide bottom, flips 180° and spins on its narrow handle. As far as I know, there may still be no consensus as to how it works.

When I think about the topic of physics 'toys', I find it striking how often the phenomena that puzzle and amuse us involve the element of spinning. As a child, I was ever entranced by a small gyroscope precariously perched at the end of my finger or horizontally suspended by a loop of string in apparent defiance of the laws of gravity. The gyroscope still fascinates me even though, as a physicist, I understand how it

---

[1] R. Boyle, *New Experiments, Physico-mechanicall, touching the Spring of the Air and its Effects* (London, 1660). A discussion of these experiments may be found in R. Harré, *Great Scientific Experiments* (Oxford, New York, 1983) 74–83.

[2] This picture is reproduced in *Niels Bohr: A Centenary Volume*, ed. A. P. French and P. J. Kennedy (Harvard University Press, Cambridge, 1985) 177.

works. My own children are intrigued by a 'one-way' spinner which I usually borrow from them for use in lectures on chiral asymmetry. It is a four-inch piece of plastic (bearing the words 'Turn on to Science') shaped like the hull of a clipper ship with just the slightest inequivalence between port and starboard sides. Spin it counterclockwise and it turns freely; spin it clockwise and it soon wobbles vehemently, stops and rotates in the opposite sense! It is startling to behold and by no means trivial to explain. In fact, the definitive explanation of its behaviour may have been provided only recently by cosmologist Hermann Bondi some hundred years after this remarkable behaviour was first reported[3]. Bondi's paper is not bedtime reading.

I have myself often fashioned a 'two-way' spinner from a wooden pencil by carving a row of notches along its length and affixing a propeller (a popsicle stick works well) to the eraser with a pin. Stroke the notches with another pencil, and the propeller spins either clockwise or counterclockwise depending on a subtle manipulation by the stroker. How does ostensibly linear motion rotate the propeller? In some way, of course, the strokes must generate elliptical vibrations in the pencil, but the details are hardly obvious.

If the motion of solids, with relatively few degrees of freedom, can be puzzling, one can only begin to imagine the paradoxical possibilities that arise when fluids are admitted. Consider, for example, the simple radiometer found in many a museum giftshop. Illuminated by bright sunlight, the four vanes (black on one side, white on the other), spin wildly about a vertical shaft inside a highly evacuated bulb. Clerk Maxwell, I have been told, was ready to discard his electromagnetic theory of light upon learning that the vanes spun the 'wrong' way – the wrong way, that is, if one assumes the vanes are driven by light pressure. It is not light, however, but residual gas that lies at the heart of the matter, although exactly how is still, more than a hundred years later, a question for discussion[4].

Some years ago, while living in Japan, my family and I encountered near a train station in the town of Hakone the eerie strains of a most unearthly symphony. There, about twenty metres in front of us, a dozen

[3] H. Bondi, The Rigid Body Dynamics of Unidirectional Spin, *Proc. R. Soc. Lond.* **A405** (1986) 265.

[4] Maxwell did not remain confused over the radiometer effect for long, but addressed its mechanisms in a seminal paper, On Stresses in Rarified Gases Arising from Inequalities of Temperature, *Phil. Trans. R. Soc. Lond.* **A170** (1879) 231, reprinted in *The Scientific Papers of James Clerk Maxwell*, Vol. 2, ed. W. D. Niven (Dover, New York, 1952) 681–712.

or so Japanese children – whose number was quickly augmented by my own – were feverishly grabbing long flexible coloured plastic tubes from the stand of a streetside vendor and twirling them furiously above their heads like lariats. The burst of tones that emerges from each musical pipe, designated 'The Voice of the Dragon' by a sign in English, soared and dropped with rotational speed over what seemed like a good portion of the range of a flute. I have never forgotten this loud, wavering, rich-toned chorus of 'dragon voices'. Despite the outward simplicity of the toy, the details of its sound production are by no means trivial, and the efforts to understand it provided both my students and me worthwhile lessons in physics as well as much entertainment.

But when it comes to gases, nowhere are the intriguing effects of rotational motion as counterintuitive, I think, as in the case of the 'Wirbelrohr'.

* * * * * * * * * * * *

I am not an avid reader of science fiction, and, in fact, except for the 'classics' by such writers as Jules Verne and H. G. Wells, generally avoid this genre of literature altogether. My father-in-law Fred, however, who for many years was a machinist at the AT&T Bell Laboratories, was a science fiction enthusiast with subscriptions to a number of such magazines spanning at least four decades. I recall in particular one visit to his home when, in the course of a chess game, we began discussing some of the strange devices he had constructed for engineers during his employment at Bell.

Perhaps the strangest device that he had ever made, however, he made for himself, he told me. I asked him what it looked like, and he replied that it was extremely simple: a hollow tube shaped like a T with no moving parts of any kind. Upon my enquiring as to what it did, Fred cocked his head, and I could see – or at least imagine I saw – a gleam in his eyes and the faint trace of a sardonic smile beneath his bushy white beard. It was a Wirbelrohr, he explained; you blew into the stem, and out one end of the cross-tube flowed hot air, while cold air flowed out the other. I laughed; I was certain he was teasing me. Although I had never heard of a Wirbelrohr, I recognised a Maxwell demon when it was described[5]. I asked my father-in-law whether he invented the device, to which Fred replied that he first read about it in one of his science fiction magazines. 'Yes indeed!' I thought to myself, and the look on my face

[5] A comprehensive guide to the literature on Maxwell demons is given by H. S. Leff and A. F. Rex, Resource Letter MD-1: Maxwell's Demon, *Am. J. Phys.* **58** (1990) 201.

undoubtedly conveyed my incredulity as if my thoughts were audible. He insisted that it worked; and when it worked really well, the cold air could freeze water, and the hot air could fry an egg!

I saw from Fred's expression that he was not teasing me. My father-in-law hailed from Switzerland; he was no physicist, but his skill in making things was exceptional. I had often thought to myself that he could make anything – although I meant, of course, anything real. Maxwell demons were, as far as I knew, imaginary. My curiosity was thoroughly aroused – all the more since I happened to be teaching a course in thermodynamics that same semester.

To my great disappointment, Fred had kept no record of the device he made, nor was he able to recall exactly when or from what magazine he obtained construction drawings. After all, he built the device some thirty years earlier. Nevertheless, having never discarded a single volume of his science fiction library, Fred promised that, as time permitted, he would search for the intriguing story. At the end of the visit, I returned home excited, but by no means convinced that the Second Law of thermodynamics should be omitted from my lectures.

Two weeks later a copy of the desired article arrived in the mail[6]. There, sandwiched between the last page of 38 000 Achnoid alien carbon people without brain chords, and the first page of encephalographic analysts led by a powerful mental mutant during the dying days of the First Galactic Empire . . . (two undoubtedly gripping tales that vividly reminded me once again why I rarely read science fiction) . . . were the anatomical details of a Maxwell demon.

My father-in-law had certainly told me the truth (as I said, he was Swiss). He, in fact, did more than that; he machined in his basement workshop a working model which I received from him shortly afterwards. The exterior was more or less just as he had described it: two identical long thin-walled tubes (the crossbar of the 'T') were connected by cylindrical collars screwed into each end of a short section of pipe that formed the central chamber; a gas inlet nozzle (the stem of the 'T'), shorter than the other two tubes but otherwise of identical construction, joined the midsection tangentially (Figure 6.1). Externally, except for a throttling valve at the far end of one output tube to control air flow, the entire device manifested bilateral symmetry with respect to a plane through the nozzle perpendicular to the cross-tubes.

Only someone with the lung capacity of Hercules could actually blow

---

[6] A. C. Parlett, Maxwell's Demon and Monsieur Ranque, *Astounding Science Fiction* (January 1950) 105–10.

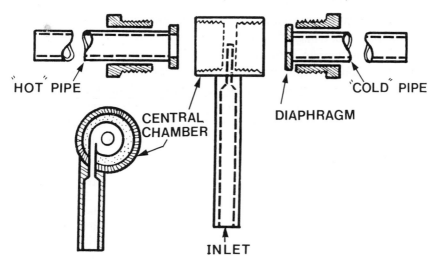

"HOT" PIPE
'COLD' PIPE
CENTRAL CHAMBER
DIAPHRAGM
INLET

Figure 6.1. Schematic diagram of a Wirbelrohr or vortex tube. Room-temperature compressed air enters the inlet tube, spirals around the central chamber, and exits through the 'hot' pipe with unconstrained cross-section or through the 'cold' pipe whose aperture is covered by a diaphragm.

into the stem. Instead, the nozzle was meant to be attached to a source of compressed air. Taking the Wirbelrohr into my laboratory, I looked sceptically for a moment at its symmetrical shape before opening the valve by my work table that started the flow of room-temperature compressed air. Then, with frost forming on the outside surface of one tube, I yelped with pain and astonishment when, touching the other tube, I burned my fingers!

* * * * * * * * * * * *

*Thermo*dynamics is different from any other dynamics in physics; in fact, the very word 'thermodynamics' is a misnomer. Whereas the term dynamics ordinarily embraces the idea of a system evolving in time under the action of specific forces (e.g. electrodynamics, hydro-dynamics, aerodynamics and chromodynamics[7]), the classical theory of thermodynamics is a study of systems in thermal equilibrium – systems, that is, whose macroscopic thermal properties are temporally unchanging. How some physical system has come to be in a state of equilibrium – or how much time is required for the system to go from one equilibrium state to another when external conditions are changed –

---

[7] Quantum chromodynamics refers to the strong interactions of elementary particles (quarks) mediated by a type of charge whimsically designated 'colour'.

is outside the principal concern of thermodynamics; for a problem in *this* area, contact your local specialist in kinetics.

To those unfamiliar with the subject, it may seem that, by excluding from its domain the intricate details of specific interactions, thermodynamics must necessarily be a weak and ineffective science compared with the other dynamical siblings in the family of physics. This, however, is not the case at all. In autobiographical notes that he merrily designated his 'obituary', Albert Einstein, a man who spent a lifetime developing physical theories, wrote[8]

A theory is the more impressive the greater the simplicity of its premises is, the more different kinds of things it relates, and the more extended is its area of applicability. Therefore the deep impression which classical thermodynamics made upon me. It is the only physical theory of universal content concerning which I am convinced that, within the framework of the applicability of its basic concepts, it will never be overthrown . . .

The strength of thermodynamics, as emphasised by Einstein, lies in its close and simple ties to experiment and observation. Let the whole edifice of chromodynamics – and therefore the theory of matter, itself – collapse because quarks turn out to be nonexistent; the conclusions of thermodynamics will remain as sound as ever.

The vast latticework of thermodynamic interrelations rests principally upon two major laws, the First and the Second. The First Law is a generalised statement of energy conservation and is to be found in one form or another in all the dynamical theories of physics. In short, energy can be transformed – from mechanical to electrical, or from electrical to heat, for example – but it cannot be created from 'nothing' or destroyed. The Second Law, however, which can be expressed in a variety of seemingly inequivalent ways, is unique to thermodynamics. In its essence, the Second Law affirms the essential irreversibility of natural or spontaneous processes.

The more tangible versions of the Second Law, attributable in one form to Lord Kelvin and Max Planck and in another to Rudolph Clausius, reveal the historical roots of thermodynamics in the practical problems of machine making. According to Kelvin–Planck no process is possible the sole result of which is to absorb heat and convert it into work. 'But what about the steam engine?' someone is bound to ask. True, it operates by heating water to steam which, upon expansion, does work; but that is not the *sole* result; heat is also discarded to the

---

[8] A. Einstein, Autobiographical Notes, in *Albert Einstein: Philosopher-Scientist*, Vol. 1, ed. P. A. Schilp (Open Court, La Salle, Illinois, 1969) 32.

environment. Neither the steam engine, nor any other engine, converts 100% of the absorbed heat into work without in some way changing the state of the rest of the world (including, possibly, itself). According to the different perspective of Clausius, no process is possible the sole result of which is to transfer heat from a cooler to a hotter body. Now, before one is tempted to assert that a refrigerator does exactly that, let him recall again the restricting condition. A refrigerator does take heat from a cool body and pumps it to a hotter body, usually the ambient air, but only upon the input of work in the form of electrical energy. The foregoing two statements of the Second Law are completely equivalent; it is not difficult to show that violation of one implies violation of the other.

Whether it is a tribute to the indomitable spirit, or simply the perversity, of human nature, the interdictions posed by the Second Law have been a red flag before the eyes of many a bullish inventor. Energy is something most people, rightly or wrongly, believe they understand at least to some degree – and the thought of building a device that generates more energy than is employed to run it is probably not seriously entertained except by those entirely ignorant of all science. The content of the Second Law, however, which embodies highly abstract notions for what is otherwise so concrete a science, rests less easily on the mind. It stands as a challenge to the ingenious as well as the ingenuous.

The frustrating thing about the Second Law is that it forbids processes for which energy remains conserved and which one might naïvely hope can be made to work – somehow. But they can't. A coin dropped from a height above a table top falls down and heats up a little upon impact. No one, I suspect, has ever witnessed a coin spontaneously rise up against the force of gravity at the expense of its own internal thermal energy thereby suffering a drop in temperature. In either case mechanical and thermal energy can be made to balance, but the process occurs in one direction only. In a similar way, the outcome of setting a hot coin on top of a cold one yields a final state of two lukewarm coins. One could wait, as they say, 'until Hell freezes over' before the time-reversed process, whereby the lower coin becomes perceptibly colder by spontaneously transferring heat to the upper coin, ever occurs, even though the total energy of the two-coin system is again unchanged. The foregoing hyperbolic remark actually serves a purpose; it emphasises an essential part of the unique quality of the Second Law *vis-à-vis* all other physical laws: its statistical validity.

Consider, for example, the First Law, the conservation of energy. Is it conceivable that, although energy appears to be conserved in processes involving macroscopic amounts of matter, that violations nevertheless occur from time to time on an atomic level? In the early 1920s, as physicists struggled to make sense of the structure of atoms and the nature of light, Niels Bohr, Hendrik Kramers and John Slater published a paper (the notorious B–K–S paper) rejecting Einstein's light-quantum hypothesis and holding to the view that the principles of energy and momentum conservation cannot be strictly applied to individual interactions[9]. Highly controversial, the B–K–S paper elicited much discussion within the physics community. Einstein and Pauli scathingly criticised it; Schrödinger, by contrast, was fascinated by it. In the end, however, the B–K–S theory was decisively refuted by experimental studies of the Compton effect, the scattering of light by free electrons. If the conservation of energy and momentum applied only to bulk matter averaged over time, and not to individual quantum processes, then there would be a non-negligible probability that an illuminated electron could recoil in any direction whatever. Within the limits of experimental precision, constrained ever more tightly by new methods and increasing advances in technology, every reliable measurement consistently revealed that individual interacting pairs of electrons and photons strictly conserved both energy and momentum[10].

That was probably the last time leading physicists seriously entertained the thought that the basic conservation principles of dynamics were only statistically valid. When in the 1930s the weak decay of elementary particles seemed to reveal violations of energy and momentum conservation, Pauli knew to look for an alternative explanation and predicted the existence of an elusive new particle, the neutrino.

Before the underlying statistical basis of thermal phenomena was clearly understood, some – Clausius, for example – regarded the Second Law to be rigorously valid in all domains of experience. The proscription that no process can, as a sole result, convert heat to work with perfect efficiency was interpreted strictly to mean *no* process ever. Perception of the Second Law as a manifestation of a law of large numbers was probably first recognised by James Clerk Maxwell, whose pioneering statistical studies of the distribution of particle velocities and

[9] N. Bohr, H. A. Kramers and J. S. Slater, The Quantum Theory of Radiation, *Phil. Mag.* **47** (1924) 785.

[10] Although quantum mechanics allows for energy and momentum *non*conserving virtual processes (such as the ephemeral creation and annihilation of particles in the vacuum) these violations occur over time intervals too short to be revealed directly by experiment; their existence is inferred from theory.

associated colligative phenomena would have marked him as a master theoretical physicist even had he never formulated the laws of electromagnetism.

Although the connection may not be obvious, the previous two formulations of the Second Law are equivalent to yet another formulation, more fundamental, in my view, as it is readily amenable to interpretation within the framework of the atomic theory of matter. This third version is expressible in terms of the abstract concept of entropy, which in addition to energy is one of the basic properties (or state variables) of an equilibrium thermodynamic state.

From the macroscopic perspective of thermodynamics, the *change* in entropy (which is what one actually measures) associated with the transformation of a system from one equilibrium state to another is related to the heat absorbed or released. From an atomic perspective, however, heat is the energy exchange between physical systems as a result of random molecular motion. In fact, the temperature of a sample of matter in thermal equilibrium with its environment is linearly proportional to, and therefore a measure of, the average molecular kinetic energy. Correspondingly, the entropy of the sample is a measure of the disorder of molecular motion; that is, the number of distinctly different ways in which the particles can be distributed over allowed quantum states and yet still give rise to the specific macroscopic properties (e.g. pressure, temperature, volume) exhibited by the sample. The higher the entropy, the greater the disorder.

Looked at statistically, then, the entropy change for a transformation between equilibrium states is a measure of the relative probability of finding the molecules of the system in the microscopic (quantum) states compatible with the final macroscopic equilibrium state of the sample compared with finding them in the microscopic states compatible with the initial equilibrium state. With this in mind, one can express the third version of the Second Law as follows: in any spontaneously occurring process, the entropy always increases, unless the process is reversible in which case the entropy change is zero.

Imagine a box divided into two sealed compartments of equal volume, one containing a gas, the other vacuum. Between the two compartments is a removable partition. When, by some external means, the partition is removed, the gas spreads into all the available volume until the gas pressure is uniform throughout the box. Wait as long as you please, the gas will never return to the original compartment. What, never? Well, hardly ever! For all the molecules to move in such a way as to recreate a vacuum in the second compartment would require a highly improbable configuration of molecular velocities. Suppose, as symmetry would suggest, the probability of finding a gas molecule in one side or the other of the original partition is $\frac{1}{2}$; then the probability that all N molecules spontaneously and independently diffuse to the same side is $(\frac{1}{2})^N$. At room temperature (20 °C) and 1 atmosphere pressure, a sample of gas initially confined to one cubic centimetre contains about 2.5×10^{19} molecules[11]. Therefore, compared with finding the gas uniformly spread throughout the entire available volume, the probability that all N molecules retreat to the initial compartment is roughly

$$P(N) \sim 1/(10^{10^{19}});$$

the denominator of this fraction is the number 1 followed by ten billion billion zeros. Technically, the probability is not exactly zero, but it is certainly indistinguishable from zero for all practical purposes.

I do not know the equation of state for Hell, but if one considers the latter 'frozen over' when, statistically speaking, not a proton remains in the observable universe[12], I can illustrate how small is the above number, or conversely how large is the reciprocal number. Contemporary theories of the elementary particles predict that baryon number is not rigorously conserved, and therefore protons should decay to positrons, among other things. Experiment currently places the proton lifetime in excess of 10^{33} years. Assuming there is any validity at all to the expectation of a finite proton lifetime, let us be generous and set it as 10^{40} years (or about 3×10^{47} seconds). Having on occasion seen the number of protons in the observable universe set at about 10^{80}, I shall again be lavish and estimate the proton count at 10^{100} (after all, what are a few zillion protons more or less?). If at some moment there are N_0 protons, the number $N(t)$ remaining at a time t later is given by the exponential decay law discussed previously in Chapter 2 (see equation (2.2b)). On average there will be one proton left after a time interval

$$t_1 = T \ln(N_0) \sim 10^{50} \text{ seconds}, \tag{6.1a}$$

---

[11] This follows from the ideal gas law: (Pressure)(Volume) = Nk(Absolute Temperature), where Boltzmann's constant k is 1.38×10^{-16} erg/(degree Kelvin), and the equilibrium conditions, expressed in suitable units, are: 1 atmosphere pressure = 10^6 dynes/cm^2, room temperature = 293 degrees Kelvin and volume = 1 cm^3.

[12] Without protons, there can be no neutrons bound in atomic nuclei; free neutrons decay (to protons, electrons and antineutrinos) with a lifetime of about 15 minutes. Positrons (from proton decay) and electrons would presumably combine and mutually annihilate. There should be nothing left, then, except electromagnetic radiation and neutrinos. Whether hot or cold, such a universe ought to qualify as a hellish place.

where T is the mean proton lifetime. Wait another 10^{40} years, or a total time still on the order of 10^{50} seconds, and there is a fair chance that even that last proton will have decayed. Hell is now completely disintegrated, let alone frozen.

How long must one wait for the gas molecules to evacuate the second compartment? Let us assume – because it is simplest to do so and because other models will hardly make any difference in the final results – that the molecules can be treated as spherical balls of some specified radius and mass. To be concrete let the mass be that of the proton (1.67×10^{-24} gram) and the radius be on the order of the Bohr radius (1×10^{-8} centimetre). It follows from the kinetic theory of gases that under the equilibrium conditions of room temperature (20 °C), 1 atmosphere pressure, and a volume of 1 cubic centimetre, the molecules in the gas move at a mean speed of about 10^5 centimetres/second and undergo roughly 10^{28} collisions per second[13]. A complete rearrangement of the approximately 10^{19} molecular velocities should therefore occur about once every 10^{-9} second. However, only one in $10^{10^{19}}$ rearrangements is likely to yield the desired configuration. Thus the time interval for the molecules to return to the first compartment would be

$$10^{-9} \text{ seconds} \times 10^{10^{19}} \sim 10^{10^{19}} \text{ seconds.} \tag{6.1b}$$

Note that $(10^{19} - 9)$ in the exponent is still about 10^{19}. The number $1/P(N)$ is *so* large that the time scale for complete molecular rearrangement obtained from any reasonable model of molecular collisions remains insignificant in comparison.

The time expressed in relation (6.1b) exceeds the putative lifetime of all matter in the universe (estimated in relation (6.1a)) to such an extent that the gas molecules in the container will have long since crumbled to photons and neutrinos before totally evacuating the second compartment. (Actually, the container, itself, would no longer exist.)

By contrast, if the box originally contained only one molecule, the likelihood of 'all' the gas being found in the original compartment is clearly 50%, and one could expect this configuration to recur over the time interval required for the molecule to traverse the length of the container, namely in about $(1 \text{ cm})/(10^5 \text{ cm/s})$ or 10 microseconds.

---

[13] The mean speed v (technically the root mean-square speed) of molecules of mass M can be estimated by equating the mean molecular kinetic energy $\frac{1}{2}Mv^2$ and mean thermal energy $\frac{3}{2}kT$; thus $v \sim (kT/M)^{1/2}$. A single molecule of cross-sectional area A sweeps out a volume Av per second of travel within which occur about $N(Av/V)$ collisions with other molecules in the container of volume V. Thus, the total rate at which molecular collisions occurs is approximately N^2Av/V per second.

Looked at from the perspective of probability, the Second Law represents, not an absolute interdiction, but rather a continuum of possibilities. When few particles are involved, the behaviour of the system is invariant under time reversal – that is, processes can occur in either direction – in keeping with the fundamental equations of motion (such as Newton's second law or the equations of Schrödinger and Dirac) that do not distinguish an 'arrow' of time. When, however, the numbers of particles involved are unimaginably huge, the spontaneous transformation of a system proceeds in that direction for which the resulting molecular configuration is overwhelmingly probable – the direction in which entropy increases.

Having understood the statistical nature – and wishing to illustrate the limitations – of the Second Law, Maxwell, noted for his incisive intellect and playful spirit, proposed a mechanism that has since become an integral part of thermodynamic lore[14]:

[The Second Law] ... is undoubtedly true as long as we can deal with bodies only in mass, and have no power of perceiving or handling the separate molecules of which they are made up. But if we conceive a being whose faculties are so sharpened that he can follow every molecule in its course, such a being ... would be able to do what is at present impossible to us.

And so was born the famous (or perhaps infamous) Maxwell demon. What could such a demon do?

Now let us suppose that ... a vessel is divided into two portions, A and B, by a division in which there is a small hole, and that a being, who can see the individual molecules, opens and closes this hole, so as to allow only the swifter molecules to pass from A to B, and only the slower ones to pass from B to A. He will thus, without expenditure of work raise the temperature of B and lower that of A, in contradiction to the second law of thermodynamics.

At the time of its enunciation in the early 1870s (at the end of an elementary textbook on heat), Maxwell's little 'being' elicited little interest in several of the major thermodynamicists then alive. Clausius responded that the Second Law did not concern what heat could do with the aid of demons, but rather what it could do by itself. Ludwig Boltzmann, who contended with Clausius for priority in deriving the Second Law from mechanics, also side-stepped the problem by arguing that in the absence of all temperature differences characteristic of thermal equilibrium, no intelligent beings could form. But to discard Maxwell's

---

[14] J. C. Maxwell, *Theory of Heat*, 8th Ed. (Longmans, Green and Co., London, 1885) 328–9.

demon as merely frivolous is to miss an essential point seized upon by later physicists; namely, whether or not an intelligent intervention (not necessarily a demon's) can exploit in some way the naturally occurring thermodynamic fluctuations within a system to circumvent the Second Law.

By about 1914 it was already quite clear that no inanimate mechanism could do this. Although phenomena such as Brownian motion and critical opalescence showed clearly that substantial fluctuations in the thermodynamic properties of bulk matter in thermal equilibrium can be made to occur[15], such fluctuations would also affect any mechanism devised to operate Maxwell's 'trap door' in a way that admitted or rejected molecules selectively. Moreover, the smaller the mechanism, the stronger would thermal fluctuations act upon it, and correspondingly the more uncontrollable would be the outcome.

The final loophole, however, that of a device operated by intelligent beings, was eliminated by the nuclear physicist Leo Szilard whose broad interests also embraced major contemporary issues in the life sciences. In what is now regarded as a classic paper[16] relating the concepts of physical entropy and information, Szilard argued that any intelligent being, even a demon, would have to make a measurement of some kind in order to exploit naturally occurring fluctuations; the very act of measuring would result in an entropy production sufficient to prevent violation of the Second Law. The idea was carried further some twenty years later when Leon Brillouin[17] demonstrated more concretely that a Maxwellian demon, working in an isolated system in thermal equilibrium, could not see the molecules. Bathed in a surrounding sea of isotropic blackbody radiation, the demon could never distinguish one molecule from another without recourse to his own source of illumination – and this additional light would generate an increase in entropy.

All of this, of course, has not ended discussion of Maxwell's demon.

---

[15] Brownian motion refers to the random movement of small particles in a fluid, for example pollen grains in water, as a result of the spatially nonuniform impacts by the molecules of fluid. Critical opalescence is the onset of a milky appearance in an initially transparent fluid at a temperature and pressure close to those for which a phase change occurs. Large fluctuations in the density, and therefore in the refractive index, of the fluid lead to substantial light scattering at all wavelengths; hence the whitish appearance.

[16] L. Szilard, Ueber die Entropieverminderung in einem thermodynamischen System bei Eingriffen intelligenter Wesen, *Zeits. Physik* **53** (1929) 840. The paper is reprinted together with an English translation in *The Collected Works of Leo Szilard: Scientific Papers*, ed. B. T. Feld and G. W. Szilard (MIT Press, Cambridge, 1972) 103.

[17] L. Brillouin, 'Maxwell's Demon Cannot Operate: Information and Entropy. I, *J. Appl. Phys.* **22** (1951) 334.

Nevertheless, from the time of Maxwell's proposal around 1871 to the present, no one has ever found or constructed a functioning demon, and it is probably accurate to state, as did Nobel laureate thermodynamicist Percy Bridgman, 'that the entire invention of the demon is most obviously a paper and pencil affair'[18].

So, what about the Wirbelrohr?

* * * * * * * * * * * *

I withdrew my fingers quickly, shut off the air supply and stared anew at my father-in-law's present. When frost at the cold end melted and the temperature of the hot end dropped, I dismantled the device, half expecting to see some diabolical little creature inside smiling at me. Actually, it was clear at the outset that the Wirbelrohr could never have functioned as a Maxwell demon. The demon effected a temperature difference in bulk matter without performing work, i.e. in violation of the Second Law. The mere fact that the Wirbelrohr had to be fed compressed air signified that initial work was done on the gas. Nevertheless, how the Wirbelrohr converted work into such a striking difference in temperature was a mystery to me.

With the few parts of the Wirbelrohr laid out on my table, I understood better the significance of the German name, 'Wirbelrohr', or vortex tube. The heart of the device is the central chamber with a spiral cavity and offset nozzle. Compressed gas entering this chamber streams around the walls of the cavity in a high-speed vortex. But what gives rise to spatially separated air currents at different temperatures? Regarding the pieces closely, I recognised immediately what had hitherto escaped my attention when I had only the story from the science fiction magazine (which scarcely made an impression on me as long as I thought it could be a hoax). Although there were indeed no moving parts of any kind, the internal geometry of the device belied the outward bilateral symmetry. The symmetry was broken by the placement in one cross-tube of a small-aperture diaphragm that effectively blocked the efflux of gas along the walls of the tube thereby forcing this part of the air flow to exit through the other arm whose cross-section was unconstrained.

The glimmer of a potential mechanism dawned on me. Had the incoming air conserved angular momentum, the rotational frequency of air molecules nearest the axis of the central chamber would be higher – as would also be the corresponding rotational kinetic energy – than

---

[18] P. W. Bridgman, *The Nature of Thermodynamics* (Harvard University Press, Cambridge, Mass., 1941) 161.

peripheral layers of air. However, internal friction between gas layers comprising the vortex would tend to establish a constant angular velocity throughout the cross-section of the chamber. In other words, each layer of gas within the vortex would exert a tangential force upon the next outer layer, thereby doing work upon it at the expense of its internal energy (while at the same time receiving kinetic energy from the preceding inner layer). Energy would consequently flow from the centre radially outward to the walls generating a system with a low-pressure, cooled axial region and a high-pressure, heated circumferential region. Because of the diaphragm, the cooler axial air had to exit one tube (the cold side), whereas a mixture of axial and peripheral air exited the other (the hot side).

The presence of the throttling valve on the hot side now made sense. If the low pressure of the air nearest the axis of the tube fell below atmospheric pressure, the cold air would not exit at all; instead, ambient air would be sucked *into* the cold end – which is what I found to be the case when the valve was fully open. By throttling the flow, pressure within the central chamber was increased sufficiently so that air could exit both tubes. Thus this simple, yet ingenious, device transferred energy within its working fluid (air) by means of a mechanism incorporating no moving parts except for the fluid itself.

But even if no demon was at work, did the Wirbelrohr violate – or come close to violating – the Second Law? Since it involved a complex, turbulent fluid flow, the operation of an actual vortex tube could not be described strictly by thermodynamics alone. Nevertheless, with some simplifying assumptions I was able to calculate the entropy change incurred by passage through the Wirbelrohr of a fixed quantity of gas of known initial temperature and pressure. Under what is termed adiabatic conditions – i.e. with no heat exchange with the environment – the Second Law requires that the entropy change of the gas, alone, be greater than or equal to zero. The resulting mathematical expression, augmented by the equation of state of an ideal diatomic gas and the conservation of energy (First Law of thermodynamics), yields an inequality

$$x^f[(1 - fx)/(1 - f)]^{1-f} \geq (p_f/p_i)^{2/7} \qquad (x = T_c/T_i) \qquad (6.2)$$

relating the temperature (T_c) of the cold air flow to the initial temperature (T_i) and pressure (p_i) of the compressed air, the fraction (f) of gas directed through the cold side, and the final pressure (p_f) of the ambient gas (taken to be 1 atmosphere). From the First Law the temperature (T_h) of the hot air flow can be expressed in terms of T_c and T_i.

By setting the expression for the entropy change equal to zero, I could calculate the lowest temperature that the cold tube should be able to reach if the gas flow were an ideal reversible process. The result was astonishing. With an input pressure of 10 atmospheres and the throttling valve set for a fraction $f = 0.3$, compressed air at room temperature (20 °C) could in principle be cooled to about -258 °C, a mere 15 degrees above the absolute zero of temperature! (The corresponding temperature of the hot side would have been 80 °C.) Clearly, the actual performance of the vortex tube, whose operation was by no means a reversible process, was far from any limitation posed by the Second Law. That did not make it any the less fascinating.

Intrigued to know more about the tube, I returned to the obviously nonfiction science fiction article that Fred sent me and tracked down the couple of references provided therein. The first experimental demonstration of a vortex tube seems to have been reported in 1933 by a French engineer, Georges Ranque[19]. Since, at the time, the device was the subject of a patent application, Ranque provided no drawings or quantitative analysis. Nevertheless, I was pleased to find from the general principles he enunciated that I had arrived at a broad explanation largely coincident with his own.

Little more was apparently heard of this device until about thirteen years later when, after the Second World War, detailed experimental investigations of German physicist Rudolph Hilsch came to the attention of an American chemist, R. M. Milton, of Johns Hopkins University who had Hilsch's work published in English[20]. In Hilsch's hands, proper selection of the air fraction f (approximately $\frac{1}{3}$) and an input pressure of a few atmospheres gave rise to an amazing output of 200 °C at the hot end and -50 °C at the cold end. Hilsch, who was the one (not my father-in-law) to coin the term 'Wirbelrohr', used the tube in place of an ammonia precooling apparatus in a machine to liquefy air.

What alerted the writer of the science fiction article to the existence of Hilsch's work was an initially brief report in the news section of an

[19] G. Ranque, Expériences sur la Détente Giratoire avec Productions Simultanées d'un Echappement d'air Chaud et d'un Echappement d'air Froid, *J. de Physique et Radium* **4**(7) (1933) 112 S.

[20] R. Hilsch, The Use of the Expansion of Gases in a Centrifugal Field as Cooling Process, *Rev. Sci. Instrum.* **18** (1947) 108; translation of an article in *Zeit. Naturwis.* **1** (1946) 208.

American chemical engineering journal[21] in 1946. The information was apparently furnished by Milton who had visited Hilsch's laboratory and brought back (or perhaps constructed later) a small model of the vortex tube. Milton, according to the journal report, was not satisfied with the interpretation of Hilsch and Ranque that frictional loss of kinetic energy produced the radial temperature distribution. Upon requesting journal readers to submit their own interpretations, the reporter apparently found himself inundated by a flood of letters from all over the world, a few excerpts of which appeared in a second report, also in 1946. Then, signing off cheerily with the hope that the information might provide a solid basis for further investigation, the reporter ceased all mention – as far as I knew – of the vortex tube.

Left with a farrago of explanations and a slim collection of old references, I looked wistfully at my Wirbelrohr. Did anyone really know how it worked?

****

Faced with other more pressing matters, I put the tube aside except for occasional classroom demonstrations. Some time afterwards, when the Wirbelrohr was all but forgotten, I experienced one of those little serendipitous twists of fate that make life interesting. Standing in a corridor of a convention centre and biding my time between sessions that interested me at a physics conference, I scanned a pile of papers strewn over a near-by table. Suddenly, one of the papers, an abstract of a talk to be given (or quite possibly already given) at a different scientific society than the one then convening, caught my eye; in its title I saw the words 'Ranque–Hilsch Effect'. Upon returning home, I wrote to the first author of the abstract who kindly sent me a copy of his papers[22]. What was proposed therein, supported by experiment, was a mechanism far different from anything that I had seen proposed before.

According to a story often told in connection with Wolfgang Pauli, an eccentric genius whose acerbic criticism could be devastating (he was

---

[21] R. L. Kenyon, Maxwellian Demon at Work, *Indus. Eng. Chem.* **38**(5) (1946) 5; The Demon Again, *ibid.* (12) 5–14.

[22] M. Kurosaka, Acoustic Streaming in Swirling Flow and the Ranque–Hilsch (Vortex Tube) Effect, *J. Fluid Mech.* **124** (1982) 139; M. Kurosaka, J. Q. Chu and J. R. Goodman, *Ranque–Hilsch Effect Revisited: Temperature Separation Traced to Orderly Spinning Waves or 'Vortex Whistle'*, presented at a conference of the American Institute of Aeronautics and Astronautics in 1982.

called 'Die Rache Gottes' – the wrath of God), Pauli presented a new theory of elementary particles before an audience including Niels Bohr. Bohr, by contrast known for his gentle qualities, could nevertheless rise to the occasion when critical remarks were required. 'We are all agreed that your theory is crazy', he allegedly replied. 'The question which divides us is whether it is crazy enough to have a chance of being correct. My own feeling is that it is not crazy enough'[23]. While the paper that I read was certainly not crazy, it seemed to me sufficiently strange and original to have a chance of being correct.

With a loud roar air rushes turbulently through the Wirbelrohr, just as it does through a jet engine or a vacuum cleaner. Buried within that roar, however, is a pure tone, a 'vortex whistle' as it has been called, that emerges from the selective amplification of background noise. Although high-pitch whistles are often associated with the swirling flow of gas in turbomachinery with rotating shafts and blades, the vortex whistle can be produced as well by the tangential introduction and swirling of gas in a stationary tube. It is this pure tone or whistle, whose frequency increases with the velocity of swirling – and hence with the pressure of the compressed air – that is purportedly responsible for the spectacular separation of temperature in a vortex tube.

The Ranque–Hilsch effect is a steady-state phenomenon – i.e. an effect that survives averaging over time. How can a high-pitch whistle – a sound that, depending on air velocity and cavity geometry, can be on the order of a few kilohertz – influence the steady (or, in electrical terms, the d.c.) component of flow? The answer, so the authors contended, was by 'acoustic streaming'. As a result of a small nonlinear convection term in the fluid equation of motion, an acoustic wave can act back upon the steady flow and modify its properties substantially. In the absence of unsteady disturbances, the air flows in a 'free' vortex around the axis of the tube; the speed of the air is close to zero at the centre (like the eye of a hurricane), increases to a maximum at around mid-radius, and drops to a small value near the walls of the tube. Acoustic streaming, however, deforms the free vortex into a 'forced' vortex within which the air speed increases linearly from the centre to the periphery. Acoustic streaming and the production of a forced vortex, rather than mere static centrifugation, engender the Ranque–Hilsch effect.

The experimental test of this hypothesis could not have been any

[23] W. H. Cropper, *The Quantum Physicists* (Oxford, New York, 1970) 57.

more direct. Remove the whistle – and *only* the whistle – and see whether the radial temperature distribution remains. To do this, the authors first monitored the entire roar with a microphone and sent the resulting electrical signal to a signal analyser that decomposed it into composite frequencies of which the discrete component of lowest frequency and largest amplitude was identified as the vortex whistle. Next, they enclosed the central cavity of the Wirbelrohr inside a tunable acoustic suppressor: a cylindrical section of Teflon with radially drilled holes serving as acoustic cavities distributed uniformly around the circumference. Inside each hole was a small tuning rod that could be inserted fully (i.e. until it touched the outer shell of the Wirbelrohr) to close off the cavity, or withdrawn incrementally to make the cavity resonant at the specified frequency to be suppressed.

To simplify their experimental test, the authors sealed off one output of the vortex tube and monitored with thermocouples the temperature difference between the centre and periphery of the cavity (which was effectively equivalent to monitoring the temperature difference between the two output tubes). In the absence of the suppressor, an increase in the pressure of the compressed air produced, as I had myself noticed when experimenting with my own vortex tube, a louder roar and greater temperature difference. When, however, the acoustic cavity was adjusted to suppress only the frequency of the vortex whistle (leaving unaffected the rest of the turbulent roar), the temperature *difference* plunged precipitously at the instant the corresponding input air pressure was reached (Figure 6.2). In one such trial, the centreline temperature jumped a total of 33 °C from − 50 °C to − 17 °C. With further increase in air pressure, the frequency of the whistle rose and, as it exceeded the narrow band of the acoustic suppressor, the temperature difference began to increase again.

Additional evidence came from a striking transformation in the nature of the flow, itself, discernible with a touch of the hand. Before the frequency of the vortex whistle was suppressed – and while, therefore, a significant radial temperature separation was produced in the tube – the exhaust air swirled rapidly near and outside the tube periphery in the manner expected for a forced vortex. Upon suppression of the whistle, however, the forced vortex was also abruptly suppressed; now quiescent at the periphery, the air rushed out close to the centreline.

So the 'demon' in the Wirbelrohr did not merely roar – it whistled, blowing hot and cold air simultaneously out different sides of its mouth.

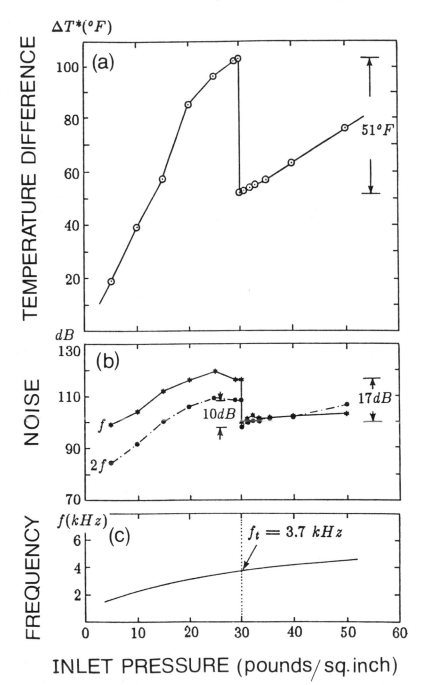

Thermodynamic analysis has shown that the Ranque–Hilsch effect is not a particularly efficient one for producing cold air. I have estimated the coefficient of performance, defined as the amount of heat removed from a mole of gas divided by the amount of work done on the gas, to be ideally just under unity, e.g. about 0.9, whereas the corresponding performance of a reversible machine (a Carnot heat pump) operating under the same conditions is about 5. The actual performance of Hilsch's tubes ran at or below 0.2. But who could be so crass as to talk about the efficiency of a remarkable phenomenon?

For all I know, the case of the mysterious Wirbelrohr is largely closed although, science being what it is, future versions of that device may yet hold some surprises in store. I have sometimes wondered, for example, what would result from supplying a vortex tube, not with room-temperature air, but with a quantum fluid, like liquid helium, free of viscosity and friction.

The exorcism of the demon in the Wirbelrohr will not, I suspect, dampen one bit the ardour of those whose passion it is to challenge the Second Law. Despite the time and effort that has been frittered away in the past, others will undoubtedly try again. On the whole such schemes are bound to fail, but every so often, as in the case of Maxwell's own whimsical creation, this failure has its positive side: when, from the clash between human ingenuity and the laws of nature, there emerge sounder knowledge and deeper understanding.

Figure 6.2. Relation of the vortex whistle to the Ranque–Hilsch effect. Increasing the pressure of the compressed air raises the frequency (f) of the vortex whistle (graph (c)). When the frequency of the whistle corresponds to the tuned frequency (f_t) of the acoustic suppressor, the contribution of the whistle (and its first harmonic at $2f$) to the acoustic noise is greatly diminished (graph (b)), the temperature at the centreline of the central chamber jumps upward, and the temperature difference between what is effectively the 'hot' and 'cold' pipes falls precipitously (graph (a)). The temperature difference grows again as f is made greater than f_t. (M. Kurosaka *et al.*, Paper AIAA-82-0952, AIAA/ASME 3rd Joint Thermophysics Conference (June 1982).)

7

Science and wonder

In the course of the more than twenty-five years that I have been studying physics, I have never found the subject dull, nor exhausted the multifarious store of interesting questions to investigate. Maybe it is lack of imagination on my part, but I have rarely felt the desire to change fields, for I could think of nothing else as deeply satisfying. Poincaré has aptly depicted the source of such feeling when he wrote (with perhaps only a modicum of exaggeration)[1]

The Scientist does not study nature because it is useful to do so. He studies it because he takes pleasure in it; and he takes pleasure in it because it is beautiful. If nature were not beautiful, it would not be worth knowing and life would not be worth living...

The contrast between a working physicist's perception of physics and the attitude held by most other people is, in my experience, simply astonishing. I have often noted with amusement the reactions shown by strangers who, upon making my acquaintance, learn what I do for a living. Following almost ritual complimentary remarks, uttered ostensibly in admiration of an intellect that can master so difficult a subject, inevitably comes a confession that physics was the speaker's most difficult and least enjoyable subject in school. If the conversation continues long enough, however, it is usually disclosed that this discomfiture originated less in difficulty than in boredom – and I learn again that the awe elicited by my occupation owes less to any presumed intellectual abilities than to the 'Sitzfleish' required to survive memorising disconnected facts and formulas and working through countless hours of tedious exercises. For *that*, unfortunately, is the impression widely left by cour-

---

[1] Cited by W. N. Lipscomb in *The Aesthetic Dimension of Science*, ed. D. W. Curtin (Philosophical Library, New York, 1980) 7.

242

ses that introduce (and often enough terminate) the study of physics.

Yet it is a fact that, throughout my entire career as a professional physicist, I have almost never had to memorise facts and formulas, but simply avail myself of the appropriate references, or, when necessary, derive the mathematical expressions I needed. What calculations I have performed – and many were indeed lengthy and time-consuming – were voluntarily undertaken to explore the topics that interested me. How can it be that a subject which provides continued intellectual challenge and pleasure to one person is at the same time the epitome of tedium and pointlessness to so many others? Anyone who teaches physics may have had occasion to think about this question. The answer, in fact, is obvious: the way in which one encounters physics in school usually bears little resemblance to the activities of a working physicist, and, to varying degrees, this may be true of other sciences as well.

* * * * * * * * * * * *

I am in the main a physicist, not a philosopher. Yet an important component of both occupations, I believe, is ably addressed in philosopher Alan Watt's self-characterisation[2]

A philosopher, which is what I am supposed to be, is a sort of intellectual yokel who gapes and stares at what sensible people take for granted, a person who cannot get rid of the feeling that the barest facts of everyday life are unbelievablly odd. As Aristotle put it, the beginning of philosophy is wonder.

Generally speaking, one might assert as well that the beginning of science is wonder. In a way, all healthy children are naturally born scientists; they come into this world with an innate and intense desire to investigate the world around them. Parents of young children know only too well (and perhaps to their frequent frustration) how difficult it is to prevent a child from exercising this inborn drive. What has happened to so many people along their paths from infancy to adulthood to so dull c cripple the natural inclination to explore and understand? Educati' theorist Jerome Bruner gets to the heart of the matter clearly[3]: *id its*

The will to learn is an intrinsic motive, one that finds both its so' *under* reward in its own exercise. The will to learn becomes a 'proble*n* is set, specialized circumstances like those of a school, where a c*in learning* students confined, and a path fixed. The problem exists not *t the natural* itself, but in the fact that what the school imposes often fai'

Press, Cambridge.

[2] A. Watts, *Does It Matter?* (Random House, New York, 1'
[3] J. S. Bruner, *Toward a Theory of Instruction* (Harvar'
1966) 115.

energies that sustain spontaneous learning – curiosity, a desire for competence, aspiration to emulate a model, and a deep-sensed commitment to the web of social reciprocity.

Human curiosity lies at the root of all science and must be nourished, or at least not thwarted, if it is to thrive. Yet science education, particularly at the introductory level, often amounts to conveying so much information that it is overwhelming, or so little that it is uninformative. The recipients of such an education can only conclude that science is mysterious and unapproachable; that scientific explanations either do not exist or cannot be understood. Then the desire to learn science is lost, perhaps irretrievably. The paramount task of science education, as I see it, is to provide inspiration, and not merely information.

There is a common theme that echoes repeatedly in the responses of scientists whenever they are asked what it is that led them to their careers. What comes across, at least in the aggregate of many such enquiries[4], is the sentiment that science is intellectually exciting, a challenge to one's mental and physical skills; that there is great beauty to science, whether in a bold and artful experimental solution to a seemingly insurmountable problem or in the aesthetic appeal and startling predictive powers of a set of equations; that there is a particular satisfaction in operating daily with universal laws and principles. Unfortunately, this does not seem to come across in the classroom. I have known many people with experiences similar to those expressed by palaeontologist Kevin Padian[5]:

...I had a fascination with dinosaurs, but science teaching deadened that fascination by [stressing prematurely] the mindless details of plant cell structures and the genetic code. It was, after all, the Age of Sputnik.

My first semester in college I [nearly failed] a required science course. I was all set for another [low grade] the second semester when the professor paused in the middle of a particularly boring lecture and said, 'You know, some of you may not be into this'. (It was now the Age of Aquarius.) 'If you'd like to do mething else, see me after class.' I did. He put me in touch with a professor tspecialised in all the things I'd always wanted to learn about, who took the ˑ help me to study them independently and discuss them.

˙d to my eventual career which I wouldn't trade for anything.

In a

ˑr way, my own educational development was more often

4 See, fo.
   1984), a.
   ber/Octob P. Weintraub, *The OMNI Interviews* (Ticknor and Fields, New York,
5 Quotation ˑle Seventy-five Reasons to Become a Scientist, *Am. Sci.* **76** (September-
   previously cˉ0.
                ˑ Reason #23, page 454, of the *American Scientist* article

ses that introduce (and often enough terminate) the study of physics.

Yet it is a fact that, throughout my entire career as a professional physicist, I have almost never had to memorise facts and formulas, but simply avail myself of the appropriate references, or, when necessary, derive the mathematical expressions I needed. What calculations I have performed – and many were indeed lengthy and time-consuming – were voluntarily undertaken to explore the topics that interested me. How can it be that a subject which provides continued intellectual challenge and pleasure to one person is at the same time the epitome of tedium and pointlessness to so many others? Anyone who teaches physics may have had occasion to think about this question. The answer, in fact, is obvious: the way in which one encounters physics in school usually bears little resemblance to the activities of a working physicist, and, to varying degrees, this may be true of other sciences as well.

* * * * * * * * * * * *

I am in the main a physicist, not a philosopher. Yet an important component of both occupations, I believe, is ably addressed in philosopher Alan Watt's self-characterisation[2]

A philosopher, which is what I am supposed to be, is a sort of intellectual yokel who gapes and stares at what sensible people take for granted, a person who cannot get rid of the feeling that the barest facts of everyday life are unbelievablly odd. As Aristotle put it, the beginning of philosophy is wonder.

Generally speaking, one might assert as well that the beginning of science is wonder. In a way, all healthy children are naturally born scientists; they come into this world with an innate and intense desire to investigate the world around them. Parents of young children know only too well (and perhaps to their frequent frustration) how difficult it is to prevent a child from exercising this inborn drive. What has happened to so many people along their paths from infancy to adulthood to so dull or cripple the natural inclination to explore and understand? Educational theorist Jerome Bruner gets to the heart of the matter clearly[3]:

The will to learn is an intrinsic motive, one that finds both its source and its reward in its own exercise. The will to learn becomes a 'problem' only under specialized circumstances like those of a school, where a curriculum is set, students confined, and a path fixed. The problem exists not so much in learning itself, but in the fact that what the school imposes often fails to enlist the natural

---

[2] A. Watts, *Does It Matter?* (Random House, New York, 1970) 25.
[3] J. S. Bruner, *Toward a Theory of Instruction* (Harvard University Press, Cambridge, 1966) 115.

energies that sustain spontaneous learning – curiosity, a desire for competence, aspiration to emulate a model, and a deep-sensed commitment to the web of social reciprocity.

Human curiosity lies at the root of all science and must be nourished, or at least not thwarted, if it is to thrive. Yet science education, particularly at the introductory level, often amounts to conveying so much information that it is overwhelming, or so little that it is uninformative. The recipients of such an education can only conclude that science is mysterious and unapproachable; that scientific explanations either do not exist or cannot be understood. Then the desire to learn science is lost, perhaps irretrievably. The paramount task of science education, as I see it, is to provide inspiration, and not merely information.

There is a common theme that echoes repeatedly in the responses of scientists whenever they are asked what it is that led them to their careers. What comes across, at least in the aggregate of many such enquiries[4], is the sentiment that science is intellectually exciting, a challenge to one's mental and physical skills; that there is great beauty to science, whether in a bold and artful experimental solution to a seemingly insurmountable problem or in the aesthetic appeal and startling predictive powers of a set of equations; that there is a particular satisfaction in operating daily with universal laws and principles. Unfortunately, this does not seem to come across in the classroom. I have known many people with experiences similar to those expressed by palaeontologist Kevin Padian[5]:

...I had a fascination with dinosaurs, but science teaching deadened that fascination by [stressing prematurely] the mindless details of plant cell structures and the genetic code. It was, after all, the Age of Sputnik.

My first semester in college I [nearly failed] a required science course. I was all set for another [low grade] the second semester when the professor paused in the middle of a particularly boring lecture and said, 'You know, some of you may not be into this'. (It was now the Age of Aquarius.) 'If you'd like to do something else, see me after class.' I did. He put me in touch with a professor who specialised in all the things I'd always wanted to learn about, who took the time to help me to study them independently and discuss them.

That led to my eventual career which I wouldn't trade for anything.

In a similar way, my own educational development was more often

---

[4] See, for example, P. Weintraub, *The OMNI Interviews* (Ticknor and Fields, New York, 1984), and the article Seventy-five Reasons to Become a Scientist, *Am. Sci.* **76** (September/October 1988) 450.

[5] Quotation taken from Reason #23, page 454, of the *American Scientist* article previously cited.

furthered by self-study or through the sympathetic counsel of those who shared their expertise with me, than through anything that occurred in the classroom.

I realise that the responsibility facing most science educators is not necessarily how to produce more scientists, but rather how to improve the quality of science instruction and to raise the level of scientific awareness of all those who would study science, even briefly. Nevertheless, it is my firm conviction that any person, whether interested in a science career or not, will be motivated to learn science for the same reasons that motivate scientists themselves: to satisfy their curiosity; to seek answers to personally meaningful questions.

If my own experiences are in any way typical, then the tragedy confronting science teachers today is that past a certain stage of development too few students retain enough curiosity about nature to ask themselves any questions. And so conventional science instruction tells its beneficiaries what precisely they must know, by when they must know it, and how they must demonstrate on tests, homework and labwork that they really know what they are supposed to. John Holt, an advocate of more self-directed education, has created a particularly appropriate simile to describe this type of teaching and learning: too often teaching is like pouring liquid into bottles that come down a conveyor belt. Questions about education reduce principally to questions about what, and how much, to pour into the bottles – two semesters of introductory physics or four? – ignoring the lack of correspondence between inanimate receptacles waiting to be filled and real people who need to make sense of the world.

* * * * * * * * * * * *

As a research physicist, I know that to perform my work I must have a sound background of factual information. But I also know why I need that information: to solve the problems that interest me. The key to successful science education must lie first in kindling curiosity where the spark has died and nurturing it where it lies latent. Only then will learning become a pleasure, and the mastery of theoretical concepts and empirical detail follow almost of their own accord.

Having taught for many years under the 'specialised circumstances' referred to by Bruner, I understand only too well how deeply ingrained in traditional educational practice these circumstances are, and how unlikely it is that they will be changed any time soon. How, then, are science teachers to sustain students' curiosity when they must do this under the restrictive conditions of parcelled time and fixed curricula that contribute in the first place to its diminution?

Faced with an educational framework over which they may have but little control, educators can nevertheless exert considerable influence through their own attitudes towards their subject and their students. There are important ties between the perception of science and the teaching of science that affect whether a teacher will be a wellspring or dry well of inspiration. Who can doubt, for instance, that a teacher for whom science is largely a technical discipline will provide a different type of instruction than one for whom science is more broadly construed as a cultural activity? Or that a teacher for whom the primary goal of science is the acquisition of accurate data will provide a different type of instruction than a teacher for whom the goal of science is the development of comprehensive theories? Attitudinal differences on the part of an educator may produce students who are interested in science for different reasons, or whose working styles, should they become scientists, are different. Such differences can be important, yet educationally innocuous.

Far more serious in their educational consequences are perspectives that seriously misconstrue the essence of science. A teacher who sees science in terms of authority figures – the allegedly great and wise who pass down their knowledge to mere mortals – may well teach science in an authoritarian manner, emphasise and require memorisation of material that scientists, themselves, would generally look up in reference books, assign problems involving needless repetition, discourage enquiries and repel with indignation challenges of fact or interpretations. A teacher who sees science as a cut-and-dried impersonal subject, a repository of facts from which correct answers inevitably flow, may well communicate to students implicitly or explicitly that human attributes and human interactions do not matter, that scientific progress follows from slavish adherence to prescribed scientific methods and not from creative imagination and resourceful use of serendipity.

By contrast, a teacher who realises that science is a multifaceted mode of enquiry and not a sepulchre of facts, that science involves personalities, and that personality and human distinctiveness affect discovery, would likely have respect for his own students' individuality, and regard that individuality as important and worth fostering. Such a teacher, even with the restrictions and inflexibility of an institutional environment, could create in the classroom an atmosphere in which students are encouraged to think, to experiment, to challenge – in short, to engage in the type of creative exploration of which science consists.

* * * * * * * * * * * *

I have come to recognise three principal tasks that science educators need to accomplish if they are to motivate the study of science.

The first is to convey an accurate and sympathetic impression of science by revealing its humanistic ties to our general intellectual heritage. In the light of heightened public concern over the impact of science on the quality of life, the physical sciences in particular are too often perceived as cold, uninspiring activities pursued by people who are at best asocial, and at worst dangerous. Norman Campbell's remarks of over seventy years ago still seem apt today[6]:

It is certain that one of the chief reasons why science has not been a popular subject ... and is scarcely recognized even yet as a necessary element of any complete education, is the impression that science is in some way less human than other studies.

If science is to be seen as a human endeavour, a quest by people for answers to significant questions, then science educators ought to provide some sense of historical perspective. Newton's laws of motion and law of gravity, for example, are among those enduring topics that will forever grace the introductory physics curriculum. Over the decades many a student, having retired for the night glassy eyed from calculating the paths of falling projectiles, must certainly have wondered, 'Why bother?' I have found, however, that when students understand better the circumstances of Newton's discoveries – that Newton addressed 'the great unanswered question confronting natural philosophy' of his time[7]; that the answers did not fall to him with ease, but only after the intense labour of 'a man transported outside himself'; that his answers had momentous impact on his contemporaries; that, being of a jealous and suspicious disposition, he had to be coaxed, flattered and wheedled by Edmund Halley (of comet fame) into writing the *Principia*; and that even Newton had trouble with the concepts of circular motion (he was, after all, inventing these concepts, not reading them from a textbook) – they look with renewed interest upon the subject. Students can be helped to realise that the laws of motion and of gravity are not artifi-

---

[6] N. Campbell, *What Is Science?* (Dover, New York, 1921) 27.
[7] Quotations are from R. Westfall, *Never at Rest: A Biography of Isaac Newton* (Cambridge University Press, London, 1980), Chapter 10, 402, 406.

cially concocted academic exercises to improve proficiency in calculation; rather, they are a precious part of our cultural legacy, a historical
landmark in mankind's progress towards knowledge and truth and away
from error and ignorance.

As part of humanising science education, teachers need also to make
their students aware of the aesthetic dimension that has long been a
source of personal pleasure and intellectual stimulation to scientists.
Sometimes this beauty is explicitly visual deriving from the colour,
shape or transformation of physical systems. Sometimes it is, as Feynman says, 'a rhythm and a pattern between the phenomena of nature
which is not apparent to the eye, but only to the eye of analysis'[8]. And
sometimes it lies in the subtle intricacy or bold simplicity of an ingenious
experimental stratagem to wrest from nature her closely held secrets. In
whatever form it takes, the beauty of science is part of a vital feedback
loop of learning: it provides motivation to explore and to comprehend
while at the same time it increases with comprehension.

The second principal task is to help students appreciate that there is
survival value to the acquisition of scientific knowledge, procedure and
attitude. In an age dominated by the fruits of science and technology, a
person ignorant of the most basic scientific principles and experimental
skills is at a severe disadvantage; he is as helpless before the forces that
mould his environment as his neolithic ancestors were before lightning
and thunder. Children, I noted earlier, are born with an innate sense of
wonder; however, they experience not only intense curiosity, but also a
strong impression of the incomprehensible. The early years of childhood
have been called the 'magic years':

These are 'magic' years because [the child's] earliest conception of the world is a
magical one; he believes that his actions and thoughts can bring about events.
Later he extends this magic system and finds human attributes in natural
phenomena and sees human or super-human causes for . . . ordinary occurrences
of daily life . . . But a magic world is an unstable world, at times a spooky world,
and as the child gropes his way toward reason and an objective world he must
wrestle with the dangerous creatures of his imagination . . .[9]

It is not exclusively children, but also those without scientific knowledge and understanding, who inhabit a 'spooky world' where ordinary
(and not so ordinary) occurrences of daily life may seem threatening.
These people have no sound foundation upon which to rely to help
distinguish plausible fact from wildly improbable speculation in the

[8] R. Feynman, *The Character of Physical Law* (MIT Press, Cambridge, 1965) 13.
[9] S. Fraiberg, *The Magic Years* (Charles Scribner's Sons, New York, 1959) ix.

barrage of imminent calamities and breakthroughs that fill the news reports. They are often paralysed in frustration by the failure or malfunction of the technological devices upon which they must depend. They are prey to the influence of occultism, mysticism, extreme religious fundamentalism and bogus science. Like the world of early childhood, theirs, too, is an unstable one troubled by the dangerous creatures of their imagination.

A person who understands more clearly how science bears on his life and affects his well-being will have more incentive to study it, for the ability to acquire for oneself needed information, to assess its reliability and to draw plausible, useful conclusions gives one confidence to decide and act in a technologically complex world.

The third principal task is to provide students an opportunity to pursue, to whatever extent possible given circumstances and resources, some form of scientific research. To participate in scientific activity directly, to have occasion to utilise the facts and techniques one is learning, is in my opinion the greatest source of motivation to study science.

All science is at root an empirical activity involving the creative interaction between theory and the facts that emerge from observation and controlled experiment. Unfortunately, most students who take science courses will never understand the role, significance and procedures of the experimental aspect of science, nor ever experience the exhilaration engendered by execution of a successful, self-motivated experiment following arduous and perhaps frustrating preparatory work. Yet it is this experimental aspect of science – the planning, looking, touching, manipulating, controlling, measuring, recording, checking – this direct contact with the phenomena of nature for the purpose of satisfying one's own curiosity, that has provided many a scientist the strongest motivation and deepest satisfaction.

Scientific experimentation has almost nothing in common with instructional laboratories that provide practical exercises designed from the outset to yield clean, unambiguous data in a reasonably short time on previously well-studied phenomena with low probability of failure. Not only does such laboratory work *not* reflect what actually transpires in a research laboratory, but, worse still, it is ordinarily *assigned* work, rather than an activity that a person would willingly and enthusiastically undertake for the purpose of learning something. Even students with little scientific experience recognise the distinction between science and 'cookbook' exercises that do not inspire – or perhaps even permit innovation and that lead in the end to results of no interest to anyone outside the classroom.

An exploratory activity, on the other hand, does not have to be of momentous general significance to science as long as it is personally meaningful to the person doing it. At times I think back to the reaction of British naturalist Alfred Wallace to the discovery of a butterfly[10]

None but a naturalist can understand the intense excitement I experienced when at last I captured it. My heart began to beat violently, the blood rushed to my head, and I felt much more like fainting than I have done when in apprehension of immediate death ... so great was my excitement produced by what will appear to most people a very inadequate cause.

One of the most pleasurable scientific experiences I have had myself was when, as a child, I built a motor out of simple nails and wire. It was pleasurable in large part because I wanted to do it; had I been required to do it, the motivation would no longer have been my own, and the educational value of the project most likely would have been lost.

* * * * * * * * * * * *

Although few people would willingly admit with any pride to ignorance of their culture's literature, music and history, I have discerned over the years no such reluctance when it comes to lack of scientific knowledge. Indeed, among science educators, researchers, employers and administrators in the United States and Great Britain in particular (and the problem is no doubt even more acute in technologically developing nations) the issue of scientific literacy – however that is interpreted – has been for a number of years now a subject of wide concern. The statistical details, if one believes them, are chilling, for they indicate massive indifference to science and gross misconceptions about the most basic facts of nature. Nor, if these reports are accurate, is there reason to expect significant favourable change in the half-century to come.

Solutions that I see proposed (depending on the level of instruction) call for such remedies as extensive standardised testing, mandatory science requirements, more classroom hours per day, longer school terms and, of course, more money to pay for all this. With respect to colleges and universities, in particular, there seems to be a growing sentiment that faculty research is at variance with good instruction, for it

---

[10] Cited in W. I. B. Beveridge, *The Art of Scientific Investigation* (Vintage, New York, 1950) 192.

means less time devoted to the classroom. The presumed remedy is to increase teaching duties and thereby reduce what critics perceive to be educationally unproductive free time.

If history is any guide, remedies predicated on the belief that science instruction and scientific research are incompatible, if not mutually adversarial, and that impose additional curricular requirements and testing as the foundation of better science teaching, are bound to fail, just as they have in the past. For what the reports of scientific illiteracy dramatically show is quite simply that, where there is no interest, science cannot be taught. One does not generate interest by increasing the very activities through which interest is lost. Real learning is, like science itself, a process of discovery, and, as educational reformers have often expressed, if one wants this process to occur in a school, then one must create the conditions under which discoveries are made: leisure to think, freedom to explore.

Science educators whose idea of instruction goes no further than the textbook, whose notes have become fossilised from unvarying use, and whose concept of scientific activity is ritualised repetition of procedure cannot hope to motivate and inspire students. Teachers must, themselves, be motivated and inspired to read avidly and regularly in order to learn the lessons of the past and to keep abreast of the present and to undertake their own investigations, however modest in scope or means, in order to teach with confidence based on personal experience.

The keys to motivating science learning are all moulded from the same metal: that science instruction is most efficacious and enduring when it reflects the intrinsic activities of science itself. To teach science well, one must have the philosophical attitudes of a scientist: to see science as culturally important, technically useful and aesthetically moving; to understand that the pursuit and acquisition of scientific knowledge helps free the mind from the bondage of ignorance, superstition and prejudice; to have a driving curiosity to comprehend the reason that manifests itself in nature and to enjoy sharing this curiosity with others.

Einstein's eloquent words say it all:

The fairest thing we can experience is the mysterious. It is the fundamental emotion which stands at the cradle of true art and true science. He who knows it not and can no longer wonder, no longer feel amazement, is as good as dead, a snuffed-out candle[11].

Our task, as educators, is to light that candle.

[11] A. Einstein, *The World As I See It* (The Wisdom Library, New York, 1949) 5.

Selected papers by the author

Chapter 1

1a. Distinctive Quantum Features of Electron Intensity Correlation Interferometry, *Il Nuovo Cimento* **B97** (1987) 200.

1b. Applications of Photon Correlation Techniques to Fermions, *Photon Correlation Techniques and Applications*, Optical Society of America Proceedings, Vol. 1, ed. J. B. Abiss and A. E. Smart (OSA, Washington D.C., 1988) 26.

[The two papers discuss quantum mechanical aspects of Hanbury Brown–Twiss type experiments uniquely characteristic of charged massive particles. Correlations attributable to the spin-statistics connection, and the influence of external potentials upon these correlations, show up in the coincidence count rate at two detectors, the variance in count rate at one detector, and the conditional probability of particle detection as a function of delay time.]

2a. On the Feasibility of Observing Electron Antibunching in a Field-Emission Beam, *Physics Letters* **A120** (1987) 442.

2b. On the Feasibility of a Neutron Hanbury Brown–Twiss Experiment, *Physics Letters* **A132** (1988) 154.

[In contrast to the belief that a short coherence time and low beam degeneracy necessarily prevents the direct observation of fermion anticorrelation in a free particle beam, the first paper shows that electron antibunching may well be observable with current technology. This is not presently the case with neutrons, however, as discussed in the second paper.]

3. Gravitationally-Induced Quantum Interference Effects on Fermion Antibunching, *Physics Letters* **A122** (1987) 226.

[The effect of gravity on the quantum mechanical phase of a multi-particle system influences the way in which the particles cluster in time and space.]

4. Fermion Ensembles that Manifest Statistical Bunching, *Physics Letters* A**124** (1987) 27.

[Although it has long been thought that electrons manifest only anticorrelations as a result of Fermi–Dirac statistics, there are in principle special types of electron states, associated with the two input beams of an interferometer, that give rise to particle correlations similar to photon bunching.]

5. Second-Order Temporal and Spatial Coherence of Thermal Electrons, *Il Nuovo Cimento* B**99** (1987) 227.

[Thermal or blackbody radiation has played a seminal role in the development of quantum mechanics and is one of the most thoroughly studied systems in physics. Although often considered to be the epitome of incoherent light, blackbody radiation does exhibit interference effects (as demonstrated, for example, by the Hanbury Brown–Twiss experiments with starlight). Examination of the coherence properties of a system of thermal electrons, a fermionic analogue of blackbody radiation, shows the profound distinctions arising from quantum statistics, the spinorial character of the basic fields, and conservation of particle number.]

6. An Aharonov–Bohm Experiment with Two Solenoids and Correlated Electrons, *Physics Letters* A**148** (1990) 154.

[Two 'back-to-back' AB experiments with momentum-correlated electrons from a single source manifest strange long-range correlations characteristic of the Einstein–Podolsky–Rosen paradox.]

7. Aharonov–Bohm Effects of the Photon', *Physics Letters* A**156** (1991) 131.

[Even though it has no electric charge, a photon can theoretically interact with the vector potential field outside a region of magnetic flux as a result of quantum electrodynamic processes involving the virtual production of correlated pairs of electrons and positrons.]

Chapter 2

1. Radiofrequency Spectroscopy of Hydrogen Fine Structure in $n = 3, 4, 5$ (with F. M. Pipkin and C. W. Fabjan), *Physical Review Letters* **26** (1971) 347.

[The paper describes the experimental investigation of a broad range of fine structure states of hydrogen atoms generated by electron capture collisions of accelerated protons with gas targets. Various radio-frequency and microwave fields are employed together to suppress over-lapping transitions and allow measurement of energy intervals between selected states.]

2. Interaction of a Decaying Atom with a Linearly Polarized Oscil-lating Field (with F. M. Pipkin), *Journal of Physics B: Atomic and Molecular Physics* **5** (1972) 1844.

[A two-level quantum system interacting with an oscillating electro-magnetic field is a fundamental quantum mechanical problem of impor-tance to spectroscopy, but for which no exact analytical solution to the Schrödinger equation is known. The equation is commonly solved in the so-called 'rotating-wave' approximation which discards anti-resonant terms in the Hamiltonian – a procedure not always adequate for unstable states, particularly in the radiofrequency domain. This paper provides a more general analytical solution to the oscillating field prob-lem that is almost indistinguishable from exact numerically integrated solutions for all cases of practical interest.]

3. Observation of Fine Structure Quantum Beats Following Step-wise Excitation in Sodium D States (with S. Haroche and M. Gross), *Physical Review Letters* **33** (1974) 1063.

[The use of two synchronously pulsed dye lasers was used to prepare sodium atoms in a linear superposition of fine structure states. An interesting feature of the experiment was to demonstrate the marked effect of light polarisation on the phase of the quantum beats; this effect was employed to enhance beat contrast.]

4. On the Anomalous Fine Structure in Sodium Rydberg States, *American Journal of Physics* **48** (1980) 244.

[A simple quantum mechanical model, based on the virtual excitation of a core electron occurring together with the actual excitation of a valence electron, was developed to account in part for the anomalous reversal of the sodium $D_{3/2}$ and $D_{5/2}$ fine structure levels.]

5. General Theory of Laser-Induced Quantum Beats, Parts I and II (with S. Haroche and M. Gross), *Physical Review* A**18** (1978) 1507, 1517.

[The two papers present a comprehensive theory of quantum beats generated by pulsed-laser excitation. The first article is concerned with the excitation of an atom by a single laser in the absence of external fields; the second treats the sequential excitation of an atom by two lasers and the influence of an external static magnetic field on the ensuing beats. Of particular interest are the nonlinear effects arising from multiple interactions between the atom and light during passage of an intense laser pulse, and the marked difference in quantum beat patterns between the cases of weak and strong external magnetic field.]

6. The Curious Problem of Spinor Rotation, *European Journal of Physics* **1** (1980) 116.

[Experimental tests are discussed that address the question of whether or not the 360° rotation of a spinor wave function can be observed.]

7. The Distinguishability of 0 and 2π Rotations by Means of Quantum Interference in Atomic Fluorescence, *Journal of Physics B: Atomic and Molecular Physics* **13** (1980) 2367.

[This paper describes the details of how, by observation of quantum beats from atoms coherently excited by a pulsed laser and subsequently irradiated by a radiofrequency field, one can distinguish a cyclic transition between two atomic states from no transition at all.]

8a. Quantum Interference Test of Orbital Angular Momentum Eigenvalues Predicted for a Spinless Charged Particle in the Presence of Long-Range Magnetic Flux, *Physical Review Letters* **51** (1983) 1927.

8b. Angular Momentum and Rotational Properties of a Charged Particle Orbiting a Magnetic Flux Tube, *Fundamental Questions in Quantum Mechanics*, ed. L. M. Roth and A. Inomata (Gordon and Breach, New York, 1986) 177.

[These papers discuss the curious properties of a charged particle in rotation about an inaccessible magnetic field, such as that within a very long solenoid. An experimental test employing a split beam of charged particles was proposed to determine which set of angular momentum eigenvalues is relevant to describing the effects of rotation. The experiment can distinguish between particle paths that wind a different number of times around the solenoid.]

9. On Measurable Distinctions Between Quantum Ensembles, *New Techniques and Ideas in Quantum Measurement Theory*, ed. D. M. Greenberger, *Annals of the New York Academy of Sciences* **480** (1986) 292.

[This paper is concerned with observable differences between quantum systems in definite, although statistically distributed, eigenstates, and systems in a linear superposition of these same eigenstates.]

10. Quantum Interference in the Fluorescence from Entangled Atomic States, *Physics Letters* A**149** (1990) 413.

[The long-distance quantum beat effect is treated in detail and shown to be insensitive to atomic motion and to allow macroscopic atomic separation.]

Chapter 3

1. Relativistic Time Dilatation of Bound Muons and the Lorentz Invariance of Charge, *American Journal of Physics* **50** (1982) 251.

[This paper, upon which the present chapter is based, discusses the issues of bound particle motion, the determination of the bound muon lifetime, and the argument for charge invariance as a consequence of atomic neutrality.]

2. Zeeman Effect in Heavy Muonic Atoms, *American Journal of Physics* **51** (1982) 605.

[A muon bound to a nucleus of sufficiently large atomic number can have a classical orbit located within the nuclear interior where the electrostatic potential experienced by the muon resembles that of a harmonic oscillator. This paper discusses the energy level structure of such a muonic atom subject to electrostatic, spin–orbit and magnetic interactions.]

3. The Lifetime of the Dimuon Atom, *Il Nuovo Cimento* D**2** (1983) 848.

[The theoretical existence of exotic atoms with *two* (or more) electrons in the same orbit replaced by unstable elementary particles raises interesting questions concerning the effect of quantum statistics on particle lifetime. Would the constraints of the Pauli exclusion principle speed up, slow down or not affect at all the decay of two ground state muons? The outcome is in some ways surprising.]

Chapter 4

1. Interference Colors with Hidden Polarizers, *American Journal of Physics* **49** (1981) 881.

[The paper provides an account of the interference colours produced by birefringent cellophane with Rayleigh scattering and Brewster angle reflection serving as the 'hidden' polarisers.]

2. Investigation of Light Amplification by Enhanced Internal Reflection, Parts I and II (with R. Cybulski, Jr), *Journal of the Optical Society of America* **75** (1983) 1732, 1739.

[The theory developed in Part I and experimental tests described in Part II of light amplification by total reflection from a medium with a population inversion are shown to be in good agreement thereby confirming the controversial phenomenon of enhanced reflection.]

3a. Specular Light Scattering from a Chiral Medium, *Lettere al Nuovo Cimento* **43** (1985) 378.

3b. Reflection and Refraction at the Surface of a Chiral Medium: Comparison of Gyrotropic Constitutive Relations Invariant or Noninvariant Under a Duality Transformation, *Journal of the Optical Society of America* A**3** (1986) 830.

[In the first paper the Fresnel amplitudes for light reflected from a transparent isotropic optically active medium are derived from the two supposedly equivalent, but fundamentally different, sets of chiral material relations and are shown to lead to physically distinguishable effects. The second paper extends the theory to absorbing chiral media and traces the origin of the inequivalence to the imposition of electromagnetic boundary conditions.]

4. Effects of Circular Birefringence on Light Propagation and Reflection (with R. Sohn), *American Journal of Physics* **54** (1986) 69.

[This paper demonstrates, among other things, the perhaps surprising result that for reflection within an anisotropic chiral medium the angle of incidence need not equal the angle of reflection.]

5. Light Reflection from a Naturally Optically Active Birefringent Medium (with J. Badoz), *Journal of the Optical Society of America* A**7** (1990) 1163.

[Ordinarily, except for propagation along special directions (the optic axes) of an anisotropic optically active medium, the effects of optical activity are overwhelmed by the much stronger linear birefringence of the medium. This paper shows that the differential reflection of circularly polarised light can still be sensitive to the weak chiral interactions of the material.]

6a. Experimental Method to Detect Chiral Asymmetry in Specular Light Scattering from a Naturally Optically Active Medium (with T. Black), *Physics Letters* A**126** (1987) 171.

6b. Experimental Configurations Employing Optical Phase Modulation to Measure Chiral Asymmetries (with N. Ritchie, G. Cushman and B. Fisher), *Journal of the Optical Society of America* A**5** (1988) 1852.

[These papers describe experimental configurations employing the photoelastic modulator to measure the differential reflection of left and right circularly polarised light, as well as other optical manifestations of left–right asymmetry in chiral media.]

7. Large Enhancement of Chiral Asymmetry in Light Reflection near Critical Angle (with J. Badoz), *Optics Communications* **74** (1989) 129.

[Under conditions of total reflection the unequal scattering of left and right circularly polarised light by an optically active medium can be orders of magnitude greater than for ordinary reflection.]

8. Differential Amplification of Circularly Polarised Light by Enhanced Internal Reflection from an Active Chiral Medium, *Optics Communications* **74** (1989) 134.

[The paper demonstrates that right and left circularly polarised light can be selectively amplified by reflection from an optically active medium with a population inversion.]

9. Wave Propagation Through a Medium with Static and Dynamic Birefringence: Theory of the Photoelastic Modulator (with J. Badoz and J. C. Canit), *Journal of the Optical Society of America* A**7** (1990) 672.

[The unusual behaviour of the photoelastic modulator first exhibited in experiments designed to test the theory of light reflection from optically active materials is fully accounted for by an analysis of light propagation through a medium with nonparallel axes of static and dynamic birefringence. The paper discusses means of circumventing the effects of static birefringence and expanding the use of photoelastic modulation to novel experimental configurations.]

10. Multiple Reflection from Isotropic Chiral Media and the Enhancement of Chiral Asymmetry (with J. Badoz), *Journal of Electromagnetic Waves and Applications* **6** (1992) 587.

[This paper discusses the conditions under which multiple reflection between two parallel optically active surfaces can enhance the difference in reflectance of left and right circularly polarised light.]

11. Chiral Reflection from a Naturally Optically Active Medium (with J. Badoz and B. Briat), *Optics Letters* **17** (1992) 886–8.
[The difference in reflection of left and right circularly polarised light from a sample of naturally optically active molecules was enhanced and quantitatively observed for the first time.]

Chapter 5

1. Satellite Test of Intermediate-Range Deviation from Newton's Law of Gravity, *General Relativity and Gravitation* **19** (1987) 511.
[This paper analyses the problem of a test mass subjected to the combined influence of gravity and the 'fifth force' within a closed spherical shell in orbit.]

2. Rotational Degeneracy Breaking of Atomic Substates: A Composite Quantum System in a Noninertial Reference Frame, *General Relativity and Gravitation* **21** (1989) 517.
[It is shown in this article that the equations of motion for the centre of mass and internal coordinates of an atom undergoing rotation can be separated (as in an inertial frame), and that coupling to the rotating frame splits otherwise degenerate magnetic substates. A quantum beat experiment with atomic Rydberg states is proposed to demonstrate this rotational level splitting.]

3a. Rotationally Induced Optical Activity in Atoms, *Europhysics Letters* **9** (1989) 95.
3b. Effect of the Earth's Rotation on the Optical Properties of Atoms, *Physics Letters* **A146** (1990) 175.
[A quantum mechanical derivation is given in the first paper of rotational circular birefringence in atoms; the treatment is generalised in the second paper to include other optical effects as well expected in the case of atoms on the spinning Earth.]

4. Circular Birefringence of an Atom in Uniform Rotation: The Classical Perspective, *American Journal of Physics* **58** (1990) 310.
[The atom in a rotating reference frame is treated by classical mechanics, and a classical interpretation of rotational optical activity is given in terms of the Coriolis force. The paper also brings out explicitly the connection between the behaviour of systems in a field-free rotating

reference frame and that in an inertial reference frame with a static magnetic field.]

5a. Measurement of Hydrogen Hyperfine Splittings as a Test of Quantum Mechanics in a Noninertial Reference Frame, *Physics Letters* A**152** (1991) 133.

5b. Optical Activity Induced by Rotation of Atomic Spin, *Il Nuovo Cimento* D**14** (1992) 857.

[It is shown that the level structure (Paper 5a) and optical properties (Paper 5b) of atoms with zero internal orbital angular momentum are nevertheless affected by rotation as a result of the coupling of particle spin to the angular velocity of rotation. The hydrogen hyperfine structure, which is one of the most precisely measured of all physical quantities, provides a good system for testing these predictions.]

Chapter 6

1. The Vortex Tube: A Violation of the Second Law?, *European Journal of Physics* **3** (1982) 88.

[The paper describes the construction of the vortex tube and gives a simple thermodynamic analysis showing that temperature separations far in excess of those actually realised can be ideally generated without violating the laws of thermodynamics.]

2. Voice of the Dragon: The Rotating Corrugated Resonator (with G. M. Cushman), *European Journal of Physics* **10** (1989) 298.

[The paper recounts a theoretical and experimental study of another simple, yet remarkable, device activated by a rotational air flow producing intriguing musical tones in a corrugated tube spun about one end. The device did not originate in Japan where I first encountered it, but, unknown to me at the time of the study, had been described earlier by F. S. Crawford, Singing Musical Pipes, *Amer. J. Phys.* **42** (1974) 278.]

Chapter 7

1. Science as a Human Endeavor, *American Journal of Physics* **53** (1985) 715.

[To help develop a much more accurate and sympathetic perspective of the nature of science, the paper describes a selection of readings and topics of discussion placing emphasis on science as a cultural activity rather than as a methodological abstraction.]

2a. Two Sides of Wonder: Philosophical Keys to the Motivation of Science Learning, *Synthèse* **80** (1989) 43.

2b. Raising Questions: Philosophical Significance of Controversy in Science, *Science and Education* **1** (1992) 163.

[Much of the present chapter was adapted from these essays (particularly the first) which served as the basis of invited talks presented at the First and Second International Conferences on the History and Philosophy of Science in Science Teaching, respectively, held at Florida State University, Tallahassee, Florida, in 1989, and at Queen's University, Kingston, Ontario, in 1992. The second paper examines some historical and contemporary scientific controversies that illustrate how different the actual practice of science is from that based on widely prevailing idealisations taught in schools.]

Index